Making Foreign Direct Investment Work for Sub-Saharan Africa

DIRECTIONS IN DEVELOPMENT
Trade

Making Foreign Direct Investment Work for Sub-Saharan Africa

Local Spillovers and Competitiveness in Global Value Chains

Thomas Farole and Deborah Winkler, Editors

© 2014 International Bank for Reconstruction and Development / The World Bank
1818 H Street NW, Washington DC 20433
Telephone: 202-473-1000; Internet: www.worldbank.org

Some rights reserved

1 2 3 4   16 15 14 13

This work is a product of the staff of The World Bank with external contributions. Note that The World Bank does not necessarily own each component of the content included in the work. The World Bank therefore does not warrant that the use of the content contained in the work will not infringe on the rights of third parties. The risk of claims resulting from such infringement rests solely with you.

The findings, interpretations, and conclusions expressed in this work do not necessarily reflect the views of The World Bank, its Board of Executive Directors, or the governments they represent. The World Bank does not guarantee the accuracy of the data included in this work. The boundaries, colors, denominations, and other information shown on any map in this work do not imply any judgment on the part of The World Bank concerning the legal status of any territory or the endorsement or acceptance of such boundaries.

Nothing herein shall constitute or be considered to be a limitation upon or waiver of the privileges and immunities of The World Bank, all of which are specifically reserved.

Rights and Permissions

This work is available under the Creative Commons Attribution 3.0 Unported license (CC BY 3.0) http://creativecommons.org/licenses/by/3.0. Under the Creative Commons Attribution license, you are free to copy, distribute, transmit, and adapt this work, including for commercial purposes, under the following conditions:

Attribution—Please cite the work as follows: Farole, Thomas and Deborah Winkler, eds. 2014. *Making Foreign Direct Investment Work for Sub-Saharan Africa: Local Spillovers and Competitiveness in Global Value Chains*. Directions in Development. Washington, DC: World Bank. doi: 10.1596/978-1-4648-0126-6. License: Creative Commons Attribution CC BY 3.0

Translations—If you create a translation of this work, please add the following disclaimer along with the attribution: *This translation was not created by The World Bank and should not be considered an official World Bank translation. The World Bank shall not be liable for any content or error in this translation.*

All queries on rights and licenses should be addressed to World Bank Publications, The World Bank Group, 1818 H Street NW, Washington, DC 20433, USA; fax: 202-522-2625; e-mail: pubrights@worldbank.org.

ISBN (paper): 978-1-4648-0126-6
ISBN (electronic): 978-1-4648-0127-3
DOI: 10.1596/978-1-4648-0126-6

Cover image: © Arne Hoel / The World Bank.
Cover design: Naylor Design, Inc.

Library of Congress Cataloging-in-Publication Data

Farole, Thomas, author.
Making foreign direct investment work for Sub-Saharan Africa : local spillovers and competitiveness in global value chains / Thomas Farole and Deborah Winkler.
    1 online resource. – (Directions in development)
"This book was prepared by Thomas Farole (Senior Economist, World Bank, PRMTR) and Deborah Winkler (Consultant Economist, World Bank, PRMTR) along with a team including: Cornelia Staritz (Austrian Research Foundation for International Development); Stacey Frederick (Center on Globalization, Governance & Competitiveness at Duke University); Rupert Barnard, Michelle de Bruyn, Philippa McLaren, and Nick Kempson (Kaiser Associates Economic Development Partners)."
    Includes bibliographical references.
    Description based on print version record and CIP data provided by publisher; resource not viewed.
    ISBN 978-1-4648-0127-3 (electronic) – ISBN 978-1-4648-0126-6 (alk. paper)
    1. Investments, Foreign—Africa, Sub-Saharan. 2. Technology transfer—Africa, Sub-Saharan. 3. Africa, Sub-Saharan—Economic conditions. I. Winkler, Deborah. II. World Bank. III. Title.
    HG5822
    332.673096—dc23   2013045059

Contents

Acknowledgments xiii
About the Editors and Contributors xv
Abbreviations xvii

	Overview	1
PART 1	**Introduction**	**5**
Chapter 1	Context, Objectives, and Methodology	7
	Abstract	7
	Why Does FDI Matter for Developing Countries?	7
	Spillovers: What Do We Know Already? What Gaps Remain in Our Understanding?	10
	Introduction to This Book	14
	Notes	19
	References	19
Chapter 2	Conceptual Framework	23
	Abstract	23
	Introduction	23
	Where and How Do Spillovers Occur?	25
	The Role of Mediating Factors	31
	Conclusions	47
	Notes	48
	References	49
PART 2	**Quantitative Studies**	**57**
Chapter 3	The Role of Mediating Factors for FDI Spillovers in Developing Countries: Evidence from a Global Dataset	59
	Abstract	59
	Introduction	59
	Empirical Model	61

	Regression Results	67
	Conclusions	78
	Annex 3A	80
	Annex 3B	81
	Annex 3C	81
	Notes	82
	References	83
Chapter 4	**Determining the Nature and Extent of Spillovers: Empirical Assessment**	**87**
	Abstract	87
	Introduction	87
	Which Foreign Investor Characteristics Increase the FDI Spillover Potential?	88
	Which Absorptive Capacities Facilitate FDI Linkages?	97
	Which Factors within Transmission Channels Support FDI Spillovers?	102
	Conclusions	111
	Notes	113
	References	113
PART 3	**Sector Case Studies**	**115**
Chapter 5	**Sector Case Study: Mining**	**117**
	Abstract	117
	The Context for Mining FDI Spillovers	117
	Supply Chain Effects	125
	Labor Market Effects	146
	Demonstration, Competition, and Collaboration Effects	151
	Conclusions	156
	Notes	160
	References	160
Chapter 6	**Sector Case Study: Agribusiness**	**163**
	Abstract	163
	The Agribusiness Global Value Chain	163
	The Context for Spillovers: Commercial Production and FDI	168
	Supply Chain Effects	175
	Labor Market Effects	189
	Demonstration, Competition, and Collaboration Effects	195
	Conclusions	203
	Notes	205
	References	206

Chapter 7   Sector Case Study: Apparel   209
Abstract   209
The Apparel Global Value Chain   209
FDI in the Apparel Global Value Chain   215
The Context for Spillovers in Sub-Saharan Africa   221
Supply Chain Effects   226
Labor Market Effects   234
Technology and Knowledge Spillovers: Demonstration and Collaboration Effects   238
Conclusions   241
Notes   243
References   244

PART 4   Conclusions and Policy Implications   245

Chapter 8   Main Conclusions   247
Abstract   247
The Evidence: Does FDI Deliver Significant Spillovers in Developing Countries?   247
The Determinants: What Mediating Factors Shape the Nature and Extent of FDI Spillovers?   254
The Role of Global Value Chains   259

Chapter 9   Policy Implications   263
Abstract   263
Introduction   263
Prioritizing the Extent of Support for Spillovers   264
Cross-Cutting versus Value Chain–Specific Interventions   265
Attracting the "Right" Foreign Investors   265
Promoting FDI–Local Economy Linkages   268
Establishing an Environment That Maximizes the Absorption Potential of Local Actors   273
Institutional and Implementation Arrangements   276
Notes   279
References   279

Boxes
1.1	Why Do GVCs Matter in the Discussion of FDI Spillovers?	15
1.2	Limitations on Robustness of Survey Findings	18
5.1	Mozambique Mining Supplier Development—Soradio	131
5.2	Supplier Upgrading in Chile—Drillco Tools	133
5.3	Does Location Matter?	134
5.4	Australian Industry Participation Plans	136

5.5	Ghana Chamber of Mines	137
5.6	Newmont Ghana's Local Procurement Policy	140
5.7	Vale SA Supplier Development Program (2012)	141
5.8	Supplier Databases in Chile and Australia	142
5.9	FDI-Driven Innovation and Supplier Upgrading: Chile's World Class Supplier Development Program	144
5.10	Skills Developments for the Mining Industry in Chile	152
6.1	Unilever's Support for Distribution Systems Upgrading	180
6.2	Cocoa and Tea: Certification Examples	183
6.3	Palm Oil: Indonesia's Nucleus-Plasma Model	186
6.4	Coffee and Cocoa: Technical Support Programs	187
6.5	Coffee: Governance and Organizational Capacity Building	188
6.6	Special-Purpose Vehicles for Technical Assistance: The Source Trust (Armajaro)	189
6.7	Examples of Entrepreneurship Linked to FDI in the Agribusiness Value Chain	191
6.8	Industry-Driven Training Programs in Kenya and Ghana	194
6.9	Fresh Produce: Horticultural Practical Training Centre, Kenya	196
6.10	National Labor Skills Certification System for Horticulture, Chile	196
6.11	Demonstration Effects—Formal and Informal	197
6.12	KenyaGAP	202
7.1	The Role of Joint Ventures in Upgrading the Sri Lankan Apparel Sector	227
7.2	Subsidized Factory Shells and an Unlevel Playing Field for Local Firms in Swaziland	234

Figures

1.1	Global Growth of FDI, 1971–2011	8
2.1	The Role of Mediating Factors for FDI Spillovers: A Conceptual Framework	24
5.1	Mining Value Chain and the Main Inputs across the Chain	118
5.2	Average Annual Sales of Domestic- and Foreign-Owned Suppliers	125
5.3	Sourcing of Goods and Services by Foreign-Owned Mining Firms	127
5.4	Domestic Firms Obtaining Quality Certifications	130
5.5	Value of Domestic Firms' Output to Foreign Customers and Market Links	132
B5.2.1	Evolution of Drillco (1990–2012)	133
5.6	Domestic Firms' Sales to Foreign Firms by Type	135
B5.9.1	Vision of the World Class Supplier Model	144
5.7	Percentage of Local Employees in Foreign Firms, by Job Category	146

5.8	Importance of Obstacles to Increasing Employment of Local Staff	149
5.9	Percentage of Foreign Firms Offering Training and Number of Days per Year	149
5.10	Foreign Investor R&D Collaboration	153
6.1	Overview of Agricultural Value Chains	165
6.2	Sourcing of Goods and Services by Foreign-Owned Agriculture Firms	176
6.3	Nature of Goods and Services Sourced Locally	177
6.4	Improvements Requested by Foreign Customers before or after Signing Contracts: Percentage Requiring Improvements and Types of Improvement	179
6.5	Market Expansion by Domestic Firms Resulting from Supply to Foreign Customers	181
6.6	Percentage of Local Employees in Foreign Firms, by Job Category	190
6.7	Importance of Obstacles to Increasing Employment of Local Staff	192
6.8	Quality and Extent of Training Offered by Foreign Firms, and Effect of Labor Turnover	193
6.9	Percentage of Domestic Firms Offering Training and Number of Training Days per Year	194
6.10	Utilization of Existing and Relevant Training Facilities by Foreign Investors and Foreign Suppliers	195
6.11	Perceived Positive Impacts of Foreign Investment on Domestically Owned Firms	197
6.12	Perceived Negative Impacts of Foreign Investors on Domestically Owned Firms across Countries	198
6.13	Perceived Technology Gap between Domestic Firms and the Top Foreign-Owned Competitor	201
6.14	Percentage of Firms That Have Relationships with Other Firms Producing Similar Products	201
7.1	Apparel Supply and Value Chain	210
7.2	Types of Lead Firms in Apparel GVCs	212
7.3	Nationality and Export Destination of Foreign-Owned Apparel Firms	223
7.4	Foreign-Owned Firms' Purchases of Goods and Services: Source, Value, and Type	228
7.5	Supply Decision Makers by Value	232
7.6	Share of Expatriate Workers in Nonproduction Positions	235
7.7	Training Facilities and the Impact of Labor Turnover on Training	237
7.8	Perceived Sophistication of Production and Technology of Kenyan-Owned Apparel Firms	239
7.9	Collaboration of Foreign-Owned Firms with Other Foreign or Locally Owned Firms in the Sector	240

9.1	Policy Framework for Spillovers	264
9.2	Framework for Measuring Local Content	271

Tables

1.1	How "Spillovers" Are Defined in This Book	17
1.2	Country and Sector Coverage	17
1.3	Overview of Methodological Approach	18
3.1	FDI Spillover Potential, FDI Spillovers	68
3.2	Absorptive Capacity, FDI Spillovers	72
3.3	National Characteristics, FDI Spillovers	75
4.1	Number of Firms by Type of Firm and Sector	89
4.2	Performance Indicators, Foreign Investors versus Domestic Producers	90
4.3	Linkages, Foreign Investors versus Domestic Producers	92
4.4	Supplier Assistance, Foreign Investors versus Domestic Producers	94
4.5	Foreign Investor Characteristics, Definition	95
4.6	Premia by Foreign Investor Characteristics	96
4.7	Distribution of Suppliers by Sector	98
4.8	Definition of Supplier Characteristics	99
4.9	Supplier Premia by Absorptive Capacity	100
4.10	The Effect of Suppliers' Absorptive Capacity on Output Share to Foreign Firms	101
4.11	The Effect of Suppliers' Absorptive Capacity with Supplier Relationship of at Least Three Years on Output Share to Foreign Firms	103
4.12	Supplier Premia by Factors within Transmission Channels	104
4.13	The Effect of Factors within Transmission Channels on the Probability of Starting to Export, Probit	105
4.14	The Effect of Assistance on the Probability of Starting to Export due to Relationship with Foreign Firm, Part 1, Probit	107
4.15	The Effect of Assistance on the Probability of Starting to Export due to Relationship with Foreign Firm, Part 2, Probit	109
5.1	Drivers of Investment by Foreign Mining Companies	122
5.2	Selected Mining Companies in Ghana, Mozambique, and Chile: Local Expenditures and "Local Procurement" Definition	126
5.3	Foreign Mining Firms' Assistance to Suppliers, Ranked by Frequency of Support	129
5.4	Assistance Received by Domestically Owned Suppliers from Foreign Customers, Ranked by Frequency of Support	130
5.5	Obstacles to Sourcing from Domestic Firms	135
5.6	Drivers of Sector Linkages among Foreign Mining Companies	154
6.1	Agribusiness Value Chain Types	164
6.2	Types of Agricultural Standards	167
6.3	Key Agricultural Exports by Value, 2010	169

6.4	Ranking of Most Important Factors Driving FDI Location Decisions in the Agricultural and Agriprocessing Sector	174
6.5	Foreign Firms' Assistance to Suppliers, Ranked by Frequency of Support	178
6.6	Types of Suppliers Receiving Assistance from Foreign Customers	184
6.7	Types of Obstacles to Further Local Sourcing by Country	184
6.8	Average Percentage of Staff with Foreign Firm Experience, by Type of Domestic Firm	191
6.9	Agricultural Research Capacity in Selected Countries	203
7.1	Firm Characteristics and Needs by Functional Category	214
7.2	Top 15 Apparel Exporting Countries, 2010	216
7.3	Sub-Saharan African Apparel Exports to the World	221
7.4	Top 10 Sub-Saharan African Apparel Exporters to the United States by Year	222
7.5	Types of FDI and Spillover Potential in Kenya, Lesotho, and Swaziland	225
7.6	Potential for Supply Chain Linkages in Kenya, Lesotho, and Swaziland	231
7.7	Factors Influencing FDI Spillovers	242
7.8	Government Support to Achieve FDI Spillovers	242

Acknowledgments

This book was prepared by Thomas Farole and Deborah Winkler (World Bank, International Trade Unit) along with a team including Cornelia Staritz (Austrian Research Foundation for International Development); Stacey Frederick (Center on Globalization, Governance and Competitiveness at Duke University); and Rupert Barnard, Michelle de Bruyn, Nick Kempson, and Philippa McLaren (Kaiser Associates Economic Development Partners). Additional background papers were provided by Niels Fold and Marianne Nylandsted Larsen (University of Copenhagen).

Many thanks to others who contributed valuable comments, data, and input including Michael Engman, Gary Gereffi, Holger Görg, Torfinn Harding, Bernard Hoekman, Giuseppe Iarossi, Steven Jaffee, Beata Javorcik, Smita Kuriakose, Daniel Lederman, Yira Mascaro, Alan Moody, Minh Cong Nguyen, Ganesh Rasagam, Jose Guilherme Reis, Federica Saliola, Murat Seker, and Ben Shepherd; and to Shienny Lie, Rebecca Martin, and Marinella Yadao for administrative support. Thanks to Michael Alwan for his keen eye in editing, Stephen Pazdan and Paola Scalabrin in External and Corporate Relations, Publishing and Knowledge for their support in publishing, and Amir Fouad, who guided us safely through the entire process.

Special thanks to our peer reviewers: Roberto Echandi (Global Product Leader, Investment Policy, World Bank, Investment Climate Department), Ted Moran (Georgetown University, Peterson Institute for International Economics, and Center for Global Development), and Kristina Svensson (Senior Operations Officer, World Bank, Sustainable Energy Department, Extractive Industries).

Finally we would like to thank the many private companies, government authorities, and other stakeholders who took the time to participate in our surveys and to meet with us to share their thoughts and experiences. Without their valuable insights this book would not have been possible.

The research was carried out under the overall supervision of Mona Haddad (Sector Manager, World Bank, International Trade Unit).

We are grateful for the support of the Bank-Netherlands Partnership Program (BNPP), which provided generous financing for the research.

About the Editors and Contributors

Editors

Thomas Farole is a senior economist at the World Bank. He is the author of *Special Economic Zones in Africa* and the editor of *Special Economic Zones: Progress, Challenges, and Future Directions* (with Gokhan Akinci) and *The Internal Geography of Trade: Lagging Regions and Global Markets*. His recent articles have appeared in the *Journal of Economic Geography, Progress in Human Geography,* and *World Development*. He holds a PhD in economic geography from the London School of Economics and Political Science (LSE), a master of science degree in local economic development from the LSE, and a bachelor of science degree in economics from the Wharton School of the University of Pennsylvania.

Deborah Winkler is a consultant economist in the World Bank's International Trade Unit and also serves as a research associate with the Schwartz Center for Economic Policy Analysis at the New School for Social Research. She is the author of the books *Services Offshoring and Its Impact on the Labor Market* and *Outsourcing Economics: Global Value Chains in Capitalist Development* (with William Milberg). Her recent articles have appeared in *World Development*, the *Journal of Economic Geography*, and *World Economy* as well as in edited volumes of the World Bank, the International Labour Office and World Trade Organization, and the Oxford Handbook Series. She received her PhD in economics from Hohenheim University in Germany.

Contributors

Stacey Frederick is a research scientist at the Center on Globalization, Governance and Competitiveness at Duke University. Her research focuses on mapping and analyzing global value chains and identifying upgrading opportunities for firms and countries in a variety of industries ranging from textiles and apparel to electronics and nanotechnology. She received her PhD in textile technology management at North Carolina State University.

Kaiser Economic Development Partners provides research, strategic analysis, and implementation support in Africa to governments, donors, and industry support structures. Kaiser Economic Development Partners has completed more than 100 projects since its establishment in 1998, with outputs including economic policies, strategies, business plans, market guides, feasibility studies, implementation frameworks, and toolkits. Kaiser Economic Development Partners has significant experience in enhancing the benefits of foreign investment to local economies, in particular through local procurement.

Cornelia Staritz is senior researcher at the Austrian Research Foundation for International Development (ÖFSE) and research associate at Policy Research in International Services and Manufacturing (PRISM) at the University of Cape Town. Prior to that, she worked at the International Trade Department of the World Bank in Washington DC and the Vienna University of Economics and Business. She holds a PhD in economics from the Vienna University of Economics and Business and a PhD in economics from the New School for Social Research in New York. Her research focuses on economic development, international trade, global value chains and production networks, private sector development, and commodity-based development.

Abbreviations

ADM	Archer Daniels Midland
AGOA	African Growth and Opportunity Act
AIA	Antofagasta Industry Association
AIP	Australian Industry Participation
BRICS	Brazil, the Russian Federation, India, China, South Africa
CAD	computer-aided design
CBI	Centre for the Promotion of Imports from developing countries
CMT	cut-make-trim
COCOBOD	Ghana Cocoa Board
CORFO	Corporación de Fomento de la Producción de Chile
CSIR	Council for Scientific and Industrial Research
EIU	Economist Intelligence Unit
ENAMI	Empresa Nacional de Mineria
EPCM	engineering, procurement, and construction management
EPZ	export processing zone
EU	European Union
EurepGAP	Euro-Retailer Produce Working Group Good Agricultural Practices
FDI	foreign direct investment
FPEAK	Fresh Produce Exporters Association of Kenya
GDP	gross domestic product
GVC	global value chain
H&M	Hennes and Mauritz
HHI	Herfindahl-Hirschman Index
HSE	health, safety, and environment
IFC	International Finance Corporation (of the World Bank Group)
ISIC	International Standard Industrial Classification
IT	information technology
IV 2SLS	instrumental variables two-stage least squares
JV	joint venture

KenyaGAP	Kenya Good Agricultural Practice
KFC	Kenya Flower Council
LIC	low-income country
LP	labor productivity
M&A	merger and acquisition
MF	mediating factor
MFA	Multi-Fibre Arrangement
MNC	multinational corporation
NGO	nongovernmental organization
OBM	original brand manufacturing
OECD	Organisation for Economic Co-operation and Development
OEM	original equipment manufacturer
OLS	ordinary least squares
PDP	Programa de Desarrollo de Proveedores (Chile)
PPD	public-private dialogue
R&D	research and development
ROO	rule of origin
SEZ	special economic zone
SME	small and medium enterprise
SOE	state-owned enterprise
SSA	Sub-Saharan Africa
TRF	Tea Research Foundation of Kenya
WDI	World Development Indicators

Overview

Foreign direct investment (FDI) is becoming an increasingly significant catalyst for output and trade in developing countries, in part due to a major expansion in the scope of global value chains (GVCs). FDI delivers a number of important contributions to economic development in terms of investment, employment, and foreign exchange. However, it is FDI's spillover potential—the productivity gain resulting from the diffusion of knowledge and technology from foreign investors to local firms and workers—that is perhaps its most valuable input to long-run growth and development. Substantial research has been undertaken on the existence and direction of spillovers from FDI, but many questions remain. Moreover, there is a need to understand better the dynamics of spillovers in certain contexts, including: (a) in low-income countries (LICs), especially in Sub-Saharan Africa (SSA); (b) outside of manufacturing sectors; and (c) in the context of GVCs. This book presents the results of a study designed to address some of these questions using a combination of desk analysis and field research in eight countries (including five in SSA) over three sectors: agribusiness, apparel, and mining. We summarize the results of this analytical work and discuss the implications for policy makers hoping to harness the power of FDI for greater development outcomes.

In chapter 2, we outline a conceptual framework for exploring the determinants of spillovers from FDI and, in doing so, provide an in-depth discussion of the existing literature and empirical evidence. The framework is built around an understanding of the mediating factors that shape the nature and extent of spillovers, specifically: the spillover potential of foreign investors (particularly in the context of investments within GVCs); the absorptive capacity of local agents (firms and workers); and how these two factors interact within a specific host country institutional environment. These mediating factors are then explored within the context of three main transmission channels through which spillovers of knowledge and technology between FDI and the domestic agents may take place: through supply chain linkages; through labor markets; and through competition, demonstration, and collaboration effects.

In part 2 of the study, we carry out econometric studies to assess the determinants of spillovers of developing countries. In chapter 3, using a cross-section of more than 25,000 domestic manufacturing firms in 78 low- and middle-income countries from the World Bank's Enterprise Surveys, we assess how mediating factors influence productivity spillovers to domestic firms from FDI. We differentiate between three types of mediating factors: (a) a foreign investor's spillover potential, (b) a domestic firm's absorptive capacity, and (c) a country's institutional framework. We find that spillovers in the short term may be negative in developing countries. This is likely caused by competition effects (particularly competition for scarce skilled labor), mainly due to weak absorptive capacity in the short run. Over time, however, FDI can lead to a beneficial restructuring of the entire industry, including opportunities for better-performing local participants and local suppliers. All three mediating factors affect the extent and direction of FDI spillovers on domestic firm productivity. Among the key results we find that FDI through joint ventures, market-seeking FDI, regional (versus global) investors, and government investment in education have a positive effect on spillovers. Conversely, a high technology gap has a negative effect.

In chapter 4, using survey data collected in Chile, Ghana, Kenya, Lesotho, Mozambique, Swaziland, and Vietnam, we first evaluate whether foreign investors differ from domestic investors in terms of their potential to generate positive spillovers for local suppliers. We find that foreign investors might offer higher potential for spillovers due to stronger performance characteristics relative to domestic counterparts. However, this is mitigated by the fact that foreign investors have fewer linkages with the local economy and offer less supplier assistance. We also find that specific characteristics of foreign investors mediate both linkages and assistance extended to local suppliers. Additionally, we examine the role of suppliers' absorptive capacities in determining the intensity of their linkages with multinationals. Our results indicate that several supplier characteristics matter, but these effects also depend on the length of the supplier's relationship with an investor. Finally, we confirm the existence of positive effects from assistance on spillovers. These effects are shown across 10 different types of assistance, among which are technical audits, joint product development, and technology licensing.

In part 3 of the book we summarize the results from in-depth sector case studies in mining, agribusiness, and apparel.

Chapter 5 covers the mining sector. For many LICs, particularly in SSA, the mining sector represents one of the most crucial sources of investment and income in their economies. This sector relies heavily on foreign investment, and FDI inflows have expanded rapidly over the past decade. Linkages and spillovers from this FDI can play a critical role in ensuring that the countries benefit over the lifespan of a mining project and develop sustainable and competitive alternative sectors. This is particularly important given that mining is a nonrenewable resources sector. This case study shows that while current linkages—in supply chains, labor markets, and wider networks—remain limited, evidence suggests there is some progress being made, and scope exists for achieving deeper local economy integration, and ultimately productivity-enhancing spillovers, over time. Indeed, the Chilean case provides an example of how to leverage the potential of the mining sector, and the Ghanaian case shows that at least some African countries are making progress down that path, if still slowly. Improving the potential for spillovers requires deep efforts in building supply-side capacity (most of which are not related directly to mining). But it also requires an active, collaborative sectoral strategy, combined with a credible and efficient regulatory approach.

In chapter 6, we explore spillovers in the agribusiness sector. Investment in the agribusiness value chain offers a significant opportunity to raise productivity levels by adopting new knowledge, technology, and techniques, from farming through processing and manufacturing. Overall, the level supply chain, labor market, and other network linkages between foreign investors and local economies are relatively higher in agriculture than in other value chains, a situation driven by the fundamental requirement of sourcing domestic agricultural inputs. Yet significant differences exist across countries, particularly in the processing and manufacturing stages of the chain. Here, SSA countries appear to have less well-established linkages than is the case in Vietnam, determined mainly by having smaller and less sophisticated domestic firms as well as fewer commercial-scale farms. The increasing importance of global standards and certification appears to be a major catalyst for supporting knowledge transfer between foreign firms and domestic actors. Efforts to promote spillovers, particularly through input provision, financing, and technical support, are widespread throughout most countries. These include efforts driven by governments, foreign investors, and multiple stakeholders; many good sector-specific approaches seem to be available. However, their sustainability is less certain, underscoring the importance of complementary and cross-cutting policies to improve skills and address supply-side constraints to competitiveness in the sector.

Chapter 7 assesses the experience of linkages and spillovers in the SSA apparel sector. In the early 2000s, the introduction of the African Growth and Opportunity Act, combined with the Multi-Fibre Arrangement quotas, contributed to a boom in FDI in the African apparel sector, leading to major growth in production, exports, and jobs. The additional possibility of exploiting the spillover potential of this FDI raised significant hopes of developing a locally embedded, competitive SSA apparel export industry. Yet more than a decade later, there has been very little progress in reaching this objective, outside of Mauritius (and South Africa, where FDI did not play a major role). Despite significant investments to attract FDI through building export processing zones and offering fiscal incentives, virtually no locally owned apparel firms are exporting or subcontracting, domestic value added remains low, local participation in management is limited, and domestic suppliers are almost absent in core and even most noncore inputs. This case study explores the level and nature of spillovers, and the factors constraining them, in three of the leading SSA apparel-exporting countries—Kenya, Lesotho, and Swaziland. We find in all three countries FDI strategies that severely limit spillover potential from the start, including a concentration in low value–added activities, external control of sourcing, and reliance on expatriates in managerial and technical positions. This is aggravated by weak domestic absorptive capacity (including weak skills development and nonexistent or inadequate domestic training institutes), barriers in the domestic business climate, ineffective policies to support local small and medium enterprises, and a missing local entrepreneurial response.

Finally, part 4 summarizes the main findings and outlines its policy implications.

Chapter 8 summarizes the main findings from this research, presenting details on the determinants of linkages and spillovers in different value chain, country, and development contexts. Despite the potential and, indeed, expectations surrounding spillovers from FDI, this study finds that short-term spillovers from FDI are typically not realized in developing countries, and may not even be positive, due in part to competition over scarce skilled labor. Over time, however, FDI can lead to a beneficial restructuring of the entire industry, including opportunities for better-performing local participants and suppliers. But therein lies the quandary: in most developing countries, foreign investors offer the most valuable potential source of knowledge and technology to build this capacity. The willingness and capacity of foreign investors to support spillovers varies hugely across sectors and firms, and is shaped by the dynamics of the GVCs in which they operate. The scale of linkages between FDI and the local economy—particularly through supply chains—is clearly the starting point. From this comes the incentive for assistance, through which knowledge and technology can be transferred.

We conclude in chapter 9 by summarizing the policy implications that emerge from the findings of this research. We provide details on the role that government—along with foreign investors, the domestic private sector, and other stakeholders—can play in promoting deeper FDI integration into domestic economies, and in harnessing more effectively the potential of FDI spillovers for development. The challenge of realizing positive spillovers from FDI in developing countries is huge. Policy makers should be aware of the difficulty in achieving local economy linkages from FDI, much less spillovers. They should also recognize that many factors outside of their control will determine from the outset the scale of these opportunities and their achievability. But government policy and programs can make a difference. Government has a role to play as a provider of information, as a facilitator, and as a regulator. Indeed, all three of these are important pieces of the solution. In sum, the generation of backward linkages and local supply chains depends on creating a favorable investment climate for local firms no less than for foreign investors. A favorable climate would include access to finance and imported inputs, enforcement of contracts, reliable regulatory standards, adequate power and other infrastructure support, and adequate competition in the domestic economy. These are the necessary, although not sufficient, conditions for success. Added to these cross-cutting factors, light-handed but deliberate and well-targeted programs that work to support building capacity and competitiveness among domestic firms can increase the likelihood of positive outcomes.

PART 1

Introduction

Context, Objectives, and Methodology   7
Conceptual Framework   23

CHAPTER 1

Context, Objectives, and Methodology

Thomas Farole and Deborah Winkler

Abstract

Foreign direct investment (FDI) is becoming an increasingly significant catalyst for output and trade in developing countries, in part due to a major expansion in the scope of global value chains (GVCs). FDI delivers a number of important contributions to economic development in terms of investment, employment, and foreign exchange. However, it is FDI's spillover potential—the productivity gain resulting from the diffusion of knowledge and technology from foreign investors to local firms and workers—that is perhaps its most valuable input to long-run growth and development. While substantial empirical evidence has been amassed over the past decade on the existence and dynamics of FDI spillovers, the results are mixed—simply attracting FDI by no means guarantees that a country will benefit from spillovers.

This chapter provides an overview of the objectives of the research for which the results are presented in this book. The aim of the research is to identify the critical factors for the realization of FDI-related spillovers—including dynamic interactions between FDI and local suppliers, service providers, workers, local producers, customers, and institutions. The research involved detailed field surveys in three industries, characterized by GVCs, across eight countries, with a specific focus on low-income countries in Sub-Saharan Africa.

Why Does FDI Matter for Developing Countries?

Perhaps the most defining characteristic of the recent period of globalization has been the proliferation of global capital flows, including foreign direct investment (FDI). Between 1990 and 2011, global FDI flows expanded more than eight-fold, 250 percent faster than world gross domestic product (GDP) and more than 60 percent faster than world trade growth over this period (figure 1.1a). Foreign affiliates of multinational corporations (MNCs) now

Figure 1.1 Global Growth of FDI, 1971–2011

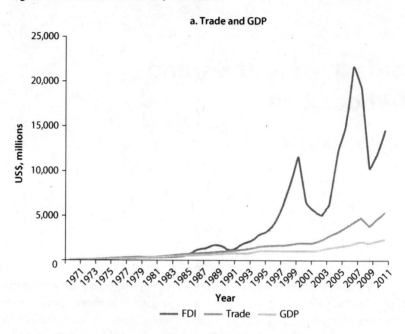

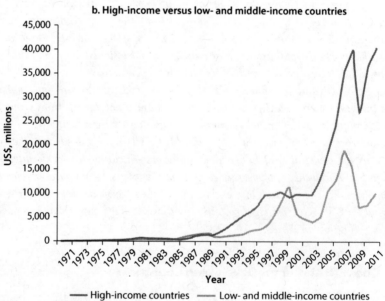

Source: World Development Indicators.
Note: FDI = foreign direct investment; GDP = gross domestic product.

employ 69 million workers and contribute US$7 million in value added (UNCTAD 2012), equivalent to more than 10 percent of all global output.

Industrialized countries still account for the majority of inward FDI stock, but the recent rapid expansion of global FDI flows has been driven by particularly strong growth of investment in developing economies (figure 1.1b). FDI inflows to low- and middle-income countries (LMICs) expanded by 30 times in just 20 years (a compound annual growth rate of 17.5 percent), almost six times faster than they did in high-income countries. As a result, the average annual share of inward global FDI flows in non–Organisation for Economic Co-operation and Development (OECD) countries rose from 16 percent during the 1970s and 1980s to reach 45 percent in 2010.[1] This trend has been supported by liberalization in global trade and investment regimes and advances in transport and communications. Together, these developments have allowed multinational firms to expand their market reach, exploit resource opportunities, and offshore activities across global production networks.

There is no debate that investment matters for economic growth. What is less clear is whether FDI is inherently more important than alternative types of investment, specifically domestic investment. Rodrik (2003) states that "[o]ne dollar of FDI is worth no more (and no less) a dollar of any other kind of investment" (cited in Moran 2005, 281). Some studies even find that the net contribution of FDI to economic growth is lower than that of domestic investment (see, for example, Firebaugh 1992 for LMICs; Tang, Selvanathan, and Selvanathan 2008 for China).

At the macroeconomic level, gains from FDI can materialize through increases in investment, employment, foreign exchange, and tax revenues (Paus and Gallagher 2008). Moreover, FDI often helps integrate host countries into the world economy (as foreign firms are engaged in exporting and use their global sales and supply networks) and, thus, stimulates trade in the long run. FDI might also have an indirect impact on skills, infrastructure, and the business environment, as countries seeking to attract foreign investment tend to put policies in place to improve these factors. Finally, foreign entry may result in more competition in the host country, leading to lower prices, more efficient resource allocation, and higher aggregate productivity (OECD 2002). At the microeconomic level, FDI can benefit local suppliers directly through increased demand for intermediates, thus raising their output, profits, and possibly investments and labor demand. FDI can also have a direct impact on domestic firms that use the output of foreign firms as inputs, possibly providing access to cheaper, higher quality, and more reliable inputs. But aside from foreign exchange and possible long-run trade stimulation, most of these benefits should also accrue from domestic investment.

So why do countries go to such lengths to attract FDI? One simple answer is that, for many developing countries, domestic capital accumulation remains too low to stimulate sufficient growth. In this context, FDI represents an important source of private capital; and given the relatively long-term outlook of direct investors (relative to portfolio investors), FDI is considered a less risky source of capital as it tends to be less vulnerable to rapid outflows in response to exogenous shocks. Another possibility is that there exist pervasive information asymmetries that result in a suboptimal level of cross-border investment.

But the main reason to attract FDI is its potential to deliver substantially greater dynamic benefits to host economies through the "spillovers" that they deliver.[2] In the context of FDI, spillovers generally refer to the diffusion of knowledge—unintentional or otherwise[3]—from multinational affiliates to domestic (local) firms. This encompasses both technology and all forms of codified and tacit knowledge related to production, including management and organizational practices. It also includes the benefits that can accrue to local actors from linking into the global networks of multinational investors. Knowledge spillovers can diffuse from foreign firms to local producers within the same industry (intra-industry or horizontal spillovers) or to another industry (inter-industry or vertical spillovers). In the latter case, they can affect local input or services suppliers in upstream sectors (backward spillovers) and local customers in downstream sectors (forward spillovers).

Spillovers: What Do We Know Already? What Gaps Remain in Our Understanding?

Given the importance of spillovers in the FDI-development nexus, what does the evidence say about the existence and nature of FDI spillovers? A substantial body of research on the existence, direction, and determinants of FDI spillovers has been built up in recent decades. Yet, while much is known about spillovers, many findings—both quantitative and qualitative—remain ambiguous, and many questions remain. Moreover, given the uncertainty of findings, research in broader contexts may be critical to gaining a deeper understanding of spillovers, and to identifying appropriate policy responses.

Snapshot of the Evidence on FDI Spillovers: Econometric Analysis and Case Studies

In recent decades, many researchers have examined econometrically the existence and direction of FDI-generated horizontal and vertical spillovers.[4] Starting with the pioneering study by Caves (1974), early research on the topic examined the effect of horizontal spillovers and generally found evidence for positive spillovers (for an overview of early studies, see, for example, Görg and Strobl 2001). Numerous studies have followed, with results showing ambiguous effects of FDI on domestic productivity. Indeed, in a meta-analysis of the spillovers literature, Görg and Greenaway (2004) suggest that the findings depend a lot on the methodology used: cross-sectional studies tend to find statistically significant evidence of positive spillovers, while results are far more ambiguous for panel studies. Similarly, Paus and Gallagher (2008) conclude in their literature overview that regressions based on cross-sectional data tend to find positive spillovers, while those based on panel data are more likely to find negative spillovers.

Shifting the focus to vertical spillovers, in a study using panel data for Lithuania, Javorcik (2004) introduces a measure of backward and forward spillovers based on input-output data, finding positive horizontal and backward spillovers, but no evidence of forward spillovers. A large body of empirical evidence has built up since this influential study. In a comprehensive meta-analysis, Havranek and Irsova (2011) take into account 3,626 estimates from 55 studies on vertical spillovers. They find evidence for positive and economically important backward spillovers from multinationals on local suppliers in upstream sectors and smaller positive effects on local customers in downstream sectors. However, the authors reject the existence of horizontal FDI spillovers.

The case study literature on FDI spillovers complements the econometric studies by providing more detail on the mechanics by which spillovers occur. This literature generally presents a more optimistic, albeit also mixed, picture, showing positive outcomes not only for vertical linkages but also through horizontal intra-industry linkages (Gallagher and Zarsky 2007) as well as labor turnover (see Blomström and Kokko 1998 for a review of older studies). Much of this work has focused on the East Asian experience. For example, Amsden and Chu (2003) analyzed partnerships between electronics MNCs and domestic firms in the 1960s and 1970s in Taiwan, China, that helped develop competitive domestic firms in the electronics sector. Focusing on vertical spillovers, Moran (1998) and Wong (2003) demonstrate how Singapore developed a network of domestic suppliers in the electronics sector based on FDI in the 1970s and 1980s. In the Malaysian electronics sector, by contrast, several studies (Ernst 2003; Phillips and Henderson 2009) depict the lack of FDI spillovers and subsequent failure to develop a deep domestic base resulting from FDI strategies that trapped domestic firms in low-value positions.

Outside Asia, studies on U.S. investment in Mexico's automotive sector in the 1980s (Moran 1998; Saggi 2002) find that FDI established extensive backward linkages and supported the upgrading of technology and global competitiveness of supplier firms. By contrast, studies in the electronics sector in Mexico, Costa Rica, and Malaysia identify a missing link between high-tech FDI and the development of domestic suppliers (Gallagher and Zarsky 2007; Paus and Gallagher 2008; Phillips and Henderson 2009), due largely to scale requirements as well as the sourcing strategies of multinationals. However, an earlier study by Larrain, Lopez-Calva, and Rodriguez-Clare (2000) finds evidence that investment by Intel in Costa Rica benefited domestic suppliers substantially through spillovers. Similar negative results are shown for the apparel sector in Sub-Saharan Africa (SSA), where more embedded Asian, European (in the case of Madagascar), and regional investors (in the case of Lesotho and Swaziland) are shown to have higher FDI spillover potential than Asian producers. The failure to develop domestic supply linkages is attributed to the latter's global sourcing strategies and a clear division of labor between global head offices—where higher-value activities and decision-making power are located—and their globally dispersed manufacturing plants (Morris, Kaplinsky, and Kaplan 2011; Morris, Staritz, and Barnes 2011; Staritz and Morris 2012, 2013).

In summary, while a vast set of empirical evidence has been amassed over the past decade on the existence and dynamics of FDI spillovers, the results are mixed, and suggest that the theoretical spillover effects often do not materialize automatically. In other words, simply attracting FDI by no means guarantees that a country will benefit from spillovers. This emphasizes the importance of understanding more clearly the conditions under which FDI can lead to positive spillovers, and what policies and institutions can contribute to establishing these conditions. In this context, however, there are significant gaps in policy-relevant research.

Remaining Research Gaps

Given the significant ambiguity in the econometric evidence on the existence and direction of spillovers as discussed, additional studies testing the existence of positive or negative spillovers are to be welcomed. Beyond this, however, the most significant scope exists for improving our understanding of the determinants and processes by which spillovers occur, and for extending the contextual field of spillovers studies. We outline six specific areas where significant research gaps remain. The area of "mediating factors" has already amassed a significant body of research, while the other five areas are much less explored.

1. *Mediating factors—identifying the conditions under which spillovers are more or less likely to materialize.* Explanations for the mixed effects of FDI spillovers include, most prominently, characteristics of foreign and domestic firms and host country factors (Castellani and Zanfei 2003; Lipsey and Sjöholm 2005). There is much recent research exploring these conditional effects, but many studies lack detail about the influence of host country characteristics and institutions on foreign and domestic firms. The FDI spillover potential of foreign investors and the development of absorptive capacities of local agents (firms and workers) are shaped by the institutional context and the policy framework in which firms operate. Mediating factors include government policies related to FDI promotion and trade, labor markets, systems of learning and innovation, finance, and taxation. Other factors include the role of institutional structures at the sectoral level, such as industry associations, employers' representatives, and trade unions, and the interactions between the public and the private sector. Several studies find evidence that a country's level of institutional development significantly affects the extent of FDI spillovers (for example, Meyer and Sinani 2009; Du, Harrison, and Jefferson 2011). But little is understood about which institutions and how. Further, international or regional regulations (for example, trade and investment policies) may affect FDI spillover potential and the transmission channels of spillovers. Understanding these institutional factors at the national, sectoral, and international/regional level is critical to formulating government policies that will stimulate FDI-generated spillovers.

2. *Transmission channels—identifying the exact mechanisms behind the observed FDI spillover patterns.* The main transmission channels through which knowledge and technology from FDI are transferred to the local

economy include supply chains, labor markets/human capital, and market forces. Within each transmission channel, FDI spillovers can be generated through several mechanisms. However, barriers within transmission channels (for example, impediments to the functioning of labor markets or competition) can interact with FDI spillovers and change their size and even direction, even if the FDI spillover potential and absorptive capacities are high. Understanding the concrete mechanisms through which FDI spillovers work and barriers or supporting factors within the different transmission channels is important to assess and increase actual spillovers.

3. ***Sectors—moving beyond manufacturing.*** Studies on FDI spillovers are highly concentrated in the manufacturing sector. This is in part because manufacturing sector data are more easily available and comparable across countries. It also reflects the large growth in FDI driven by fragmentation of global production networks, and the fact that many of the success stories of global development and competitiveness (particularly from East Asia) are linked to FDI in manufacturing supply chains. A few recent studies have explored spillovers in services industries (see, for example, Javorcik and Li 2008; Arnold, Javorcik, and Mattoo 2011). But much of the growth in FDI over the past decade, particularly in lower-income developing economies, has come through natural resources–based sectors. Such investment differs considerably from traditional manufacturing FDI. For example, investors tend to be resource seeking rather than efficiency or market seeking. Investment is also increasingly likely to be dispersed to a wider set of countries and to emerge from a widening set of investors (including large investors from the global south). Some recent studies have begun to explore the issue of linkages outside traditional manufacturing sectors, including prominent work done through the Making the Most of Commodities Programme[5] (see, for example, Morris, Kaplinsky, and Kaplan 2011).

4. ***Country context—understanding spillover dynamics in low-income countries.*** Developing countries are covered in several econometric and case studies in the FDI literature. However, work that explicitly accounts for the specific contexts of low-income countries (LICs) is limited (most studies tend to focus on middle-income East Asia and Latin America). Studies focused on LICs, particularly in regions like SSA, are limited (see, for example, Görg and Strobl 2005; Bwalya 2006; Morris, Kaplinsky, and Kaplan 2011). Nevertheless, understanding the specific spillover dynamics in these contexts is ever more important. FDI is increasingly central to the development strategies of many LICs, and they place significant emphasis on attracting it, often with substantial financial incentives and instruments like special economic zones. Such investment has been a catalyst to economic growth and development in some SSA countries. But in many other countries, FDI has failed to deliver the expected benefits and has been disconnected from the domestic economy. When developing an FDI policy agenda for

LICs, it is critical to understand how their specific institutional contexts and development concerns affect transmission channels. This knowledge will help determine the types of policies and institutions that will increase FDI spillovers.

5. *Global value chains—understanding spillover potential in the context of global strategies and governance.* Few recent studies (with notable exceptions, including Morris, Kaplinsky, and Kaplan 2011) consider the specific dynamics of global value chains (GVCs) for mediating spillovers. This is particularly important in light of recent results stressing the significance of vertical spillovers (Havranek and Irsova 2011). GVCs encompass the full range of activities that are required to bring a good or service from conception, through the different phases of production (provision of raw materials, the input of components, subassembly, producer services, and the assembly of finished goods) and delivery to final users. In the context of globalization, the activities that comprise a value chain are increasingly carried out in inter-firm networks on a global scale (see Gereffi 1999; Gereffi, Humphrey, and Sturgeon 2005). As discussed in box 1.1, the potential for and the nature of FDI spillovers may be strongly determined by the specific GVC dynamics, in terms of sectors, corporate strategies of lead firms governing GVCs, the FDI motive, and the global production and sourcing policies of foreign investors (Gallagher and Zarsky 2007; Paus and Gallagher 2008).

6. *Competitiveness—linking spillovers to exports and integration.* One of the important spillover benefits from foreign investors is their potential to help internationalize domestic firms, particularly their suppliers. They do this indirectly by bringing demands for meeting international standards (for example, in quality and delivery) and by helping build the scale and productivity of their domestic suppliers, as well as directly by providing access to their international marketing, supply, and distribution networks. While most studies of FDI spillovers focus on productivity gains to the domestic economy, there remains a gap in terms of looking explicitly at their role in facilitating export participation and establishing internationally competitive firms in host markets.

Introduction to This Book

Objectives

This book is the culmination of research that aims to contribute to the burgeoning study on foreign investment spillovers by addressing some of the policy-relevant gaps discussed in the previous section. The objective of the research is to help LICs take better advantage of the potential benefits of FDI, taking into account the specific context of GVC investment. Specifically, the research aims to identify the critical factors for the realization of FDI-related spillovers—including dynamic interactions between FDI and local suppliers,

Box 1.1 Why Do GVCs Matter in the Discussion of FDI Spillovers?

Reduced policy barriers to cross-border trade and investment, combined with substantial improvements in transport and communications technology, have allowed for significant fragmentation and geographical dispersion of production. This "second unbundling" (Baldwin 2012) has contributed to a major shift in global trade and investment patterns in recent decades, with production in individual countries increasingly forming just one stage in a product's value chain. The networks that have emerged from this process are often referred to as "global value chains" (GVCs). GVCs typically are led by large firms (normally in industrialized countries) and involve networks of suppliers across many countries in various stages (Milberg and Winkler 2013) of value addition—from raw materials through component stages and to final production, assembly, and delivery. Today, it is estimated that more than half of the value of world trade is made up of products traded in the context of GVCs (Cattaneo et al. 2013). And in low-income countries (LICs), particularly small ones, the vast majority of non-resource-seeking investment takes place in the context of GVCs.

Aside from their sheer scale, GVCs potentially have important implications for achieving spillovers:

- Joining GVCs creates significant opportunities for spillovers through rapid insertion of host countries into global networks. However, realizing and sustaining these spillover benefits may be mediated by the governance structures of value chains (Gereffi and Fernandez-Stark 2010) as well as the specific value-chain strategies of lead firms.
- Related to the above, GVC structures potentially allow developing countries to participate in global trade without building up full sector-specific supply chains (Baldwin 2012). However, these countries may have little or no domestic expertise in the value chains they are investing in, thus limiting the potential for spillovers.
- In addition, GVC-oriented investments in many cases have become enclaves, relatively disconnected from their host country economies. This has perhaps been aggravated by the tendency of GVC-oriented investment to be located within export processing zones.

Thus, while many issues related to spillovers are relevant for any type of FDI, investment in GVCs raises additional concerns. Understanding how GVCs impact spillover potential and how they differ across sectoral and locational contexts is critical for maximizing the potential benefits from FDI for LICs.

Sources: Gereffi and Fernandez-Stark 2010; Baldwin 2012; Cattaneo et al. 2013; Milberg and Winkler 2013.

---

service providers, workers, local producers, customers, and institutions. Ultimately, the research aims to:

1. Enable policy makers in LICs to adopt policies that more effectively facilitate spillovers from foreign investors to local economies.
2. Inform policy dialogue and target donor lending and technical assistance to ensure an alignment of the instruments needed to leverage FDI and maximize related spillovers.

We emphasize that this book does not aim to give guidance on the day-to-day practicalities of designing and delivering programs to support spillovers. Instead, the focus is on giving guidance on the policy environment to support spillover programs and spillovers in general over the long term.

Scope

The book is concerned with FDI spillovers—what level of spillovers is possible in different contexts and what does it take to deliver on these possibilities? Our measure of FDI spillovers is the standard one: how the presence of foreign investors affects productivity of domestic firms in the host economy. However, for the field case studies (see further discussion below), we are unable to measure productivity effects. Instead we attempt to capture the proximate outcomes and actions that are expected to contribute to spillovers, including the use of local suppliers, sales to local markets, hiring and training of workers and managers, and technical and financial assistance provided to local suppliers. In addition, we are also interested in how the presence of FDI contributes to increased exports by domestic firms (in horizontal or vertical relationships with foreign investors), as this represents an alternative measure of spillovers and can proxy for productivity gains.[6]

The concept of spillovers is fairly broad and measurement is almost always difficult. This study takes a fairly narrow approach (table 1.1) by focusing mainly on spillovers that emerge through the day-to-day interaction between foreign investors and the local economy. These interactions occur within supply chains (buying from and selling to local firms) and labor markets (employing and training workers and managers); there are also more tacit interactions that diffuse knowledge to local actors (for example, through observation of new practices and through access to knowledge and technology embedded in new products and services). Finally, the discussions in this book are clearly weighted toward supply chains. This is because substantially more quantitative and qualitative data is available for supply chain spillovers than for spillovers through either labor markets or changing market forces.

A number of other potential spillovers exist, however, which are not quantified or assessed in any detail in this study. Among the most important of these, at least in sectors like mining and agribusiness, are the infrastructure investments made by foreign investors that have the potential to improve the competitiveness of other firms and farms in the region and improve access of the local population to markets and services. On the other hand, the study also excludes discussion of other (typically negative) externalities that may result from foreign investment, including environmental effects, potential impacts on inequality, and other economic and social externalities. The main reason for excluding both of these effects was methodological (see following discussion)—measuring these effects would have required substantial resources to employ very detailed, additional methodologies beyond those used for the core of the study. Moreover, there was a risk that the analysis would get bogged down in measurement of spillovers (and cost-benefit type analysis), which is not the aim of this research. Finally, we also

Context, Objectives, and Methodology

Table 1.1 How "Spillovers" Are Defined in This Book

Spillovers included	Spillovers excluded
• Supply chain (goods and services) linkages • Employment • Knowledge, technology, practices • Networks • Changing market forces (competition and demonstration) • Exporting	• Common-use infrastructure (public or club good) • Multiplier effects from employment and linkages • Externalities (such as environment and inequality) • Forward value addition/beneficiation (limited coverage in the study)

Table 1.2 Country and Sector Coverage

	Agribusiness	Apparel	Mining
Chile			√
Ghana	√		√
Kenya	√	√	
Lesotho		√	
Mozambique	√		√
Sri Lanka		√	
Swaziland		√	
Vietnam	√		

do not analyze forward value-addition investments by FDI (although it is included in some general discussions), as these would be considered core investments rather than spillovers as such.

The study is designed to assess the experience of spillovers in LICs and has a particular focus on SSA. In addition, however, we have included analysis of countries outside the region (low and middle income) to provide additional context and, where relevant, policy lessons (good and bad) from countries that may have had a longer history of substantial foreign investment than in SSA. The extent and nature of potential FDI-generated spillovers differ importantly by sector and FDI motive. Therefore, the study attempts to cover a range of sectors that are particularly important for low-income economies, again focusing on SSA. This includes two natural resources–based sectors—mining and agribusiness (including primary and processed agriculture)—and the apparel sector (as the most important light manufacturing sector in LICs). Table 1.2 provides an overview of country and sector coverage of the study.

Methodology

The study used a combination of desk and field research, both of which included quantitative and qualitative components (table 1.3). Desk-based research included development of the study's theoretical framework (see chapter 2) from a comprehensive overview of the literature on FDI spillovers. Framework development also included an econometric analysis of FDI spillovers across more than 100 LMICs, making use of the global dataset from the World Bank's Enterprise Surveys (see chapter 3).

The core of the study, however, involved detailed field research in the eight countries outlined in table 1.2. This involved carrying out more than 400 surveys of foreign investors, domestic investors, and suppliers across the three target sectors and eight countries (as per table 1.2), making use of a survey instrument designed to assess the level and nature of interactions between investors and the local economies in which they operate. Due to lack of sector-specific data on investors and suppliers in almost all countries, sampling for the survey involved using a snowball technique to identify companies relevant across the four categories. Secondary sources such as investment promotion databases, industry association membership lists, and Internet searches were used to identify relevant firms, and personal referrals from new and existing contacts in the relevant industries were utilized to compile the sample. The vast majority of surveys were completed through face-to-face interviews. In some cases surveys were also self-completed by some of the stakeholders that met with the research team and engaged in semistructured interviews (see box 1.2).

Table 1.3 Overview of Methodological Approach

	Desk based	Field based
Quantitative	• Cross-country econometric study using World Bank Enterprise Surveys	• Analysis of results of field surveys
Qualitative	• Theoretical overview and conceptual framework	• Field surveys • Case studies

Box 1.2 Limitations on Robustness of Survey Findings

A number of limitations in the sampling methodology have been identified through the course of the research:

- *FDI and domestic firm diversity across commodities and operations.* Foreign and domestic investors were surveyed across commodities and products. This means that any comparison of competitive dynamics affecting foreign and domestic investors must be treated with caution.
- *Supplier diversity across goods and services.* Survey responses were received from firms representing a wide range of goods and services. For example, in agribusiness alone, respondents included agro-input dealers, seed distributors, equipment manufacturers, laboratory services, irrigation consultants, veterinary services, packaging suppliers, commercial farmers, and commodity traders.

Due in part to this diversity, but more importantly because the number of foreign investors was often limited to a narrow set of firms (especially in the mining sector), the total number of surveys carried out in each country is not sufficient for carrying out country-specific econometric analysis.

box continues next page

Box 1.2 Limitations on Robustness of Survey Findings *(continued)*

As a result of these limitations, the robustness of the findings from the surveys must be treated with caution. Where possible, survey findings have been supplemented with secondary research or findings from semi-structured interviews.

In addition to the surveys, more than 200 semi-structured interviews were carried out in these countries, covering these same actors as well as other stakeholders such as government ministries and agencies, support institutions, trade unions, industry bodies, and nongovernmental organizations. The results of this field research were used to carry out further quantitative analysis on the determinants of spillovers as well as to develop comprehensive case studies.

Structure of the Book

This book is structured into four parts. Part 1 provides an introduction, including this chapter as well as chapter 2, which sets out the conceptual framework for the study. Part 2 of the study covers the quantitative analytics. It is composed of two cross-country econometric studies focused on identifying the mediating factors that shape the level and nature of FDI spillovers: (a) a global study using data from the World Bank's Enterprise Surveys (chapter 3), and (b) an analysis drawing on the results from the firm surveys carried out as part of the field research (chapter 4). Part 3 summarizes the results of the sector case studies, covering mining (chapter 5), agribusiness (chapter 6), and apparel (chapter 7). Finally, part 4 concludes, providing first a summary of the main conclusions across the study (chapter 8) and a discussion of implications for policy (chapter 9).

Notes

1. Note that half of the growth in non-OECD FDI over this period is accounted for by China alone.
2. This is because multinational firms are seen to enjoy technological and other advantages that result in higher levels of productivity (Hoekman and Javorcik 2006), and that these advantages can be transferred to host country firms.
3. Technically, an intentional transfer of knowledge is considered to be a spillover only if it is not compensated in some way.
4. Most econometric studies look for the existence of spillovers by assessing whether the level or growth rate of productivity of domestic firms is higher in sectors where foreign firms are more prevalent.
5. See http://commodities.open.ac.uk/mmcp.
6. "New trade theory" (heterogeneous firms) has shown that the most productive firms in an economy tend to self-select into exporting. For a summary of the literature, see Melitz and Redding (2013).

References

Amsden, A., and W. Chu. 2003. *Beyond Late Development: Taiwan's Upgrading Policies*. Cambridge, MA: MIT Press.

Arnold, J., B. Javorcik, and A. Mattoo. 2011. "Does Services Liberalization Benefit Manufacturing Firms? Evidence from the Czech Republic." *Journal of International Economics* 85 (1): 136–46.

Baldwin, R. 2012. "Trade and Industrialisation after Globalisation's 2nd Unbundling: How Building and Joining a Supply Chain Are Different and Why It Matters." NBER Working Paper 17716, National Bureau of Economic Research (NBER), Cambridge, MA.

Blomström, M., and A. Kokko. 1998. "Multinational Corporations and Spillovers." *Journal of Economic Surveys* 12 (2): 1–31.

Bwalya, S. 2006. "Foreign Direct Investment and Technology Spillovers: Evidence from Panel Data Analysis of Manufacturing Firms in Zambia." *Journal of Development Economics* 81 (2): 514–26.

Castellani, D., and A. Zanfei. 2003. "Technology Gaps, Absorptive Capacity and the Impact of Inward Investments on Productivity of European Firms." *Economics of Innovation and New Technology* 12 (6): 555–76.

Cattaneo, O., G. Gereffi, S. Miroudot, and D. Taglioni. 2013. "Joining, Upgrading and Being Competitive in Global Value Chains: A Strategic Framework." Policy Research Working Paper 6406, World Bank, Washington, DC.

Caves, R. 1974. "Multinational Firms, Competition, and Productivity in Host-Country Markets." *Economica* 41 (162): 176–93.

Du, L., A. Harrison, and G. Jefferson. 2011. "Do Institutions Matter for FDI Spillovers? The Implications of China's 'Special Characteristics'." NBER Working Paper 16767, National Bureau of Economic Research (NBER), Cambridge, MA.

Ernst, D. 2003. "Global Production Networks in East Asia's Electronics Industry and Upgrading Perspectives in Malaysia." In *Upgrading East Asia's Industries*, edited by S. Yusuf, Vol. 2. Oxford, U.K.: Oxford University Press.

Firebaugh, G. 1992. "Growth Effects of Foreign and Domestic Investment." *American Journal of Sociology* 98 (1): 105–30.

Gallagher, K., and L. Zarsky. 2007. *The Enclave Economy: Foreign Investment and Sustainable Development in Mexico's Silicon Valley*. Cambridge, MA: MIT Press.

Gereffi, G. 1999. "International Trade and Industrial Upgrading in the Apparel Commodity Chain." *Journal of International Economics* 48 (1): 37–70.

Gereffi, G., and K. Fernandez-Stark. 2010. "The Offshore Services Value Chain: Developing Countries and the Crisis." In *Global Value Chains in a Postcrisis World: A Development Perspective*, edited by O. Cattaneo, G. Gereffi, and C. Staritz, 335–72. Washington, DC: World Bank.

Gereffi, G., J. Humphrey, and T. Sturgeon. 2005. "The Governance of Global Value Chains." *Review of International Political Economy* 12 (1): 78–104.

Görg, H., and D. Greenaway. 2004. "Much Ado about Nothing? Do Domestic Firms Really Benefit from Foreign Direct Investment?" *The World Bank Research Observer* 19 (2): 171–97.

Görg, H., and E. Strobl. 2001. "Multinational Companies and Productivity Spillovers: A Meta-Analysis." *Economic Journal* 111 (475): F723–39.

———. 2005. "Spillovers from Foreign Firms through Worker Mobility: An Empirical Investigation." *Scandinavian Journal of Economics* 107 (4): 693–709.

Havranek, T., and Z. Irsova. 2011. "Estimating Vertical Spillovers from FDI: Why Results Vary and What the True Effect Is." *Journal of International Economics* 85 (2): 234–44.

Hoekman, B., and B. Javorcik. 2006. "Lessons from Empirical Research on Technology Diffusion through Trade and Foreign Direct Investment." In *Global Integration and Technology Transfer*, edited by B. Hoekman and B. Javorcik, 1–28. Washington, DC: Palgrave/World Bank.

Javorcik, B. 2004. "Does Foreign Direct Investment Increase the Productivity of Domestic Firms? In Search of Spillovers through Backward Linkages." *American Economic Review* 94 (3): 605–27.

Javorcik, B., and Y. Li. 2008. "Do the Biggest Aisles Serve a Brighter Future? Global Retail Chains and Their Implications for Romania." Working Paper 4650, World Bank, Washington, DC.

Larrain, F., L. Lopez-Calva, and A. Rodriguez-Clare. 2000. "Intel: A Case Study of Foreign Direct Investment in Central America." CID Working Paper 58, Center for International Development, Harvard University.

Lipsey, R., and F. Sjöholm. 2005. "The Impact of Inward FDI on Host Countries: Why Such Different Answers?" In *Does Foreign Direct Investment Promote Development?* edited by T. Moran, E. Graham, and M. Blomström, 23–43. Washington, DC: Peterson Institute for International Economics and Center for Global Development.

Melitz, M., and S. Redding. 2013. "Firm Heterogeneity and Aggregate Welfare." NBER Working Paper 18919, National Bureau of Economic Research (NBER), Cambridge, MA.

Meyer, K., and E. Sinani. 2009. "When and Where Does Foreign Direct Investment Generate Positive Spillovers? A Meta-Analysis." *Journal of International Business Studies* 40 (7): 1075–94.

Milberg, W., and D. Winkler. 2013. *Outsourcing Economics: Global Value Chains in Capitalist Development*. New York: Cambridge University Press.

Moran, T. 1998. *Foreign Direct Investment and Development: The New Policy Agenda for Developing Countries and Economies in Transition*. Washington, DC: Institute for International Economics.

———. 2005. "How Does FDI Affect Host Country Development? Using Industry Case Studies to Make Reliable Generalizations." In *Does Foreign Direct Investment Promote Development?* edited by T. Moran, E. Graham, and M. Blomström, 281–313. Washington, DC: Peterson Institute for International Economics.

Morris, M., R. Kaplinsky, and D. Kaplan. 2011. "Commodities and Linkages: Meeting the Policy Challenge." MMCP Discussion Paper 14, Making the Most of Commodities Programme (MMCP), The Open University, U.K.

Morris, M., C. Staritz, and J. Barnes. 2011. "Value Chain Dynamics, Local Embeddedness, and Upgrading in the Clothing Sectors of Lesotho and Swaziland." *International Journal of Technological Learning, Innovation and Development* 4 (1–3): 96–119.

OECD (Organisation for Economic Co-operation and Development). 2002. "Foreign Direct Investment for Development, Maximising Benefits, Minimising Costs: Overview." OECD, Paris.

Paus, E., and K. Gallagher. 2008. "Missing Links: Foreign Investment and Industrial Development in Costa Rica and Mexico." *Studies of Comparative International Development* 43 (1): 53–80.

Phillips, R., and J. Henderson. 2009. "Global Production Networks and Industrial Upgrading: Negative Lessons from Malaysian Electronics." *Austrian Journal for Development Studies* 25 (2): 38–61.

Rodrik, D. 2003. Appel Inaugural Lecture, Columbia University, March 27.

Saggi, K. 2002. "Trade, Foreign Direct Investment, and International Technology Transfer: A Survey." *The World Bank Research Observer* 17 (2): 191–235.

Staritz, C., and M. Morris. 2012. "Local Embeddedness, Upgrading and Skill Development: Global Value Chains and Foreign Direct Investment in Lesotho's Apparel Industry." Working Paper 32, ÖFSE, Vienna.

———. 2013. "Local Embeddedness and Economic and Social Upgrading in Madagascar's Export Apparel Industry." Working Paper 38, ÖFSE, Vienna.

Tang, S., E. Selvanathan, and S. Selvanathan. 2008. "Foreign Direct Investment, Domestic Investment and Economic Growth in China: A Time Series Analysis." *The World Economy* 31 (10): 1292–309.

UNCTAD (United Nations Conference on Trade and Development). 2012. *World Investment Report: Towards a New Generation of Investment Policies.* Geneva, Switzerland: UNCTAD.

Wong, P. 2003. "From Using to Creating Technology: The Evolution of Singapore's National Innovation System and the Changing Role of Public Policy." In *Competitiveness, FDI and Technological Activity in East Asia*, edited by S. Lall and S. Urata, 191–238. London: Elgar.

CHAPTER 2

Conceptual Framework

Thomas Farole, Cornelia Staritz, and Deborah Winkler[1]

Abstract

We outline a conceptual framework for exploring the determinants of spillovers from foreign direct investment (FDI), and in doing so provide an in-depth discussion of the existing literature and empirical evidence. The framework is built around understanding the mediating factors that shape the nature and extent of spillovers, specifically: the spillover potential of foreign investors (particularly in the context of investments within global value chains); the absorptive capacity of local agents (firms and workers); and how these two factors interact within the institutional environment of a specific host country. These mediating factors are then explored within the context of three main transmission channels by which spillovers of knowledge and technology between FDI and the domestic agents may take place: through supply chain linkages; through labor markets; and through competition, demonstration, and collaboration effects.

Introduction

This chapter develops a conceptual framework for the study of foreign direct investment (FDI) spillovers, with a specific focus on understanding the mediating factors that shape the extent and nature of spillovers. The conceptual framework draws on the literature presented in the following sections on spillovers and mediating factors and extends the FDI spillovers model of Paus and Gallagher (2008) (see figure 2.1). The framework is built on the premise that the realization of FDI spillovers is conditioned by the *spillover potential of foreign investors*, the *absorptive capacity of local agents* (firms and workers), and the *interaction of these two factors*. This interaction is importantly determined by *host country characteristics*, the *institutional framework*, and the *transmission channels*. FDI spillover potential is a necessary condition for the realization of spillovers, but it is not a sufficient condition (Paus and Gallagher 2008).

In the next section, we assess the channels and mechanisms through which FDI spillovers can be generated, including supply chains, labor turnover, and changing market forces. In short, multinationals tend to demand higher-quality inputs, which may give local suppliers incentives to upgrade their technology, and may also diffuse knowledge to local firms. In addition, they may provide higher-quality inputs to domestic customers. Competition between local firms may increase and local firms may try to imitate the multinational's products and practices. In addition, knowledge embodied in labor can transmit from foreign to local firms through labor turnover.

These mechanisms are influenced by characteristics of foreign and domestic firms, as well as host country factors and the institutional framework, which are discussed in the third section of this chapter. Our conceptual framework in figure 2.1 depicts various mediating factors for FDI and spillovers, as identified in the literature. At the foreign investor level, factors include degree and structure of foreign ownership, FDI motive, global production and sourcing strategies in global value chains (GVCs), technology intensity, FDI home country, entry mode, and length of presence in the country. Factors at the domestic level that

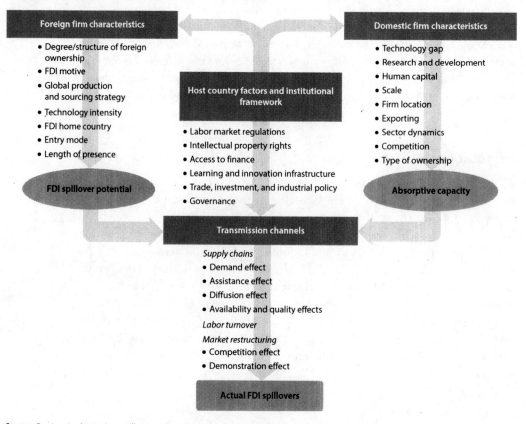

Figure 2.1 The Role of Mediating Factors for FDI Spillovers: A Conceptual Framework

Sources: Sections in chapter 2 on spillovers and mediating factors; the model of Paus and Gallagher 2008.

mediate local firms' absorptive capacity include the technology and productivity gap, research and development (R&D), human capital, firm size and scale, firm location, exporting, sector dynamics, competition, and type of ownership. Factors at the host country and institutional levels influence foreign and domestic firm characteristics and affect the transmission channels through which knowledge diffuses from multinationals to local firms. Such factors include a country's labor market regulations, intellectual property rights, access to finance, learning and innovation infrastructure, trade, investment and industrial policy, institutions and governance, and competition.

Finally, note that the processes described in this framework are interdependent. Foreign investors' production and sourcing strategies are not static. Also, changes in the characteristics of domestic firms and host countries influence these strategies in terms of local sourcing, subcontracting, and employment decisions. Hence, there is dynamic interaction between the local absorption capacity and the FDI spillover potential as indicated in figure 2.1.

Where and How Do Spillovers Occur?

Introduction

This section explores two main questions: where and how FDI spillovers occur. In the FDI literature, several channels for spillovers are identified (see, among many others, Hoekman and Javorcik 2006; Crespo and Fontoura 2007). Three main channels emerge from the research: supply chains, labor turnover, and changing market forces. Supply chain spillovers only work vertically, whereas labor turnover and changing market forces can work both horizontally and vertically. In this section, we discuss the underlying mechanisms within the transmission channels that have the potential to generate FDI spillovers. The analysis also considers factors related to foreign firms, domestic firms, and host countries that may determine the extent of FDI spillovers in specific transmission channels. These factors, summarized here, are discussed in further detail in the third section of this chapter on the role of mediating factors.

Spillovers through Supply Chains

Vertical spillovers can be categorized as backward spillovers, which occur when local firms become input or service suppliers of multinational firms, and forward spillovers, which emerge when the goods and services provided by multinational firms are used as inputs in local industries. Another spillover channel is through subcontracting linkages, where foreign firms subcontract part of their production to local firms operating in the same industry.

Demand and Assistance Effects

FDI spillovers can be generated through the demand of multinationals for better and/or more diverse inputs which, in turn, benefits all firms in the industry of the multinational. Multinationals may expect high international standards from local input and service suppliers—for example, for product quality and delivery time, and technological efficiency that increases their overall productivity (Paus and Gallagher 2008). Besides better products, foreign firms may increase the demand for more specific intermediates, which will increase input variety in upstream sectors (Javorcik 2008).

Multinational affiliates might help local producers to upgrade their technological capabilities directly through sharing of production techniques and product design and assisting with technology acquisition (Paus and Gallagher 2008). Spillovers to supplying industries may also be generated through personnel training, advance payment, leasing of machinery, provision of inputs, help with quality assurance, and organization of product lines (Lall 1980; Crespo and Fontoura 2007; Javorcik 2008). Although, in the strict sense, this mechanism is triggered by intentional knowledge transfer, it might result in real spillovers if the supplying firm doesn't fully compensate the multinational for these benefits.

The possibility for spillovers through the demand and assistance effects depends on the multinational firm's strategies regarding its FDI motives, global production and sourcing, and the relevance of its suppliers. Multinational firms that follow co-sourcing or co-location strategies, for instance, rely on established global supplier relationships, which clearly reduce the possibility and extent of vertical spillovers to local firms. Vertical spillovers are also influenced by the multinational's technology intensity. Host country characteristics and the institutional framework can shape the extent of linkages with domestic firms (for example, through requirements to use local suppliers) and the type of FDI attracted in the first place. Domestic firm characteristics, such as firm size, relative technology gap, share of R&D and skilled labor, and proximity to the multinational, among others, influence which suppliers are picked by foreign investors. These characteristics, in turn, are shaped by host country and institutional characteristics.

Diffusion Effect

Multinational assistance to a supplier firm might lead to unintentional knowledge spillovers in the supplying industry. Despite the leakage of knowledge, the multinational firm can still gain as new firm entry and increased competition among local suppliers drives down input prices. Pack and Saggi (2001) even envisage the case where the existing local supplier can benefit from this situation: If the availability of cheaper inputs results in new firm entry in the sector of the multinational, this may increase competition and input demand there and benefit all supplier firms. As long as the resulting competition from knowledge diffusion in both the supplier's and multinational's industries is not too strong, both firms can win (Javorcik 2008).

The diffusion effect can be considered an indirect spillover through changing market forces in the supplying sector. Any assistance by the foreign firm (technological or otherwise) puts domestic suppliers in an advantageous position compared to their domestic counterparts in their sector. These effects are moderated by several factors, as we discuss in more detail in the third section. Instead of foreign firm characteristics, the extent of the spillover now depends on characteristics of the local supplier receiving assistance by a foreign firm. And domestic firm characteristics now refer to other competitors in the supplying sector. Host country characteristics and the institutional framework further influence the extent of the diffusion effect.

Availability and Quality Effects

Availability and quality effects refer to the case where a multinational firm supplies to a local producer in downstream sectors. Assuming that multinational firms are more productive and technology intensive than local producers, foreign entry increases the availability, variety, and reliability of higher-quality inputs. The probability and extent of positive spillovers may be higher for services than for materials inputs for two reasons, both sector specific. First, the availability effect may be economically more important, as a larger availability of basic business services (for example, telecommunications, banking, information technology) benefits a wider range of domestic clients. Second, the performance of domestic firms in downstream sectors depends more critically on the availability and quality of service inputs, due to larger limitations on services imports (Javorcik 2008). Several factors related to the FDI motive and strategy of the multinational, characteristics of domestic client firms, and host country institutions (for example, investment promotion agencies) play a role in determining the extent of forward spillovers. However, research on mediating factors for forward spillovers is relatively scarce, as we show in the third section on mediating factors.

Empirical Evidence of Direct Supplier–Multinational Linkage Effects

Studies on FDI spillovers that focus on direct supplier–multinational linkages based on survey data are rare. Focusing on foreign affiliates in five transition economies, Giroud, Jindra, and Marek (2012) find that foreign firm characteristics have a positive impact on backward FDI linkages and spillovers. Javorcik and Spatareanu (2009) find evidence for "learning-by-supplying" for a sample of Czech manufacturing firms, although there is also evidence for self-selection into supplying due to a higher productivity ex ante. Jordaan (2011) also confirms the existence of positive backward spillovers on manufacturing suppliers in Mexico. Specifically, positive spillovers are facilitated through supplier firms' absorptive capacities and the level of support from the multinational. Studying the Polish automotive sector, Gentile-Lüdecke and Giroud (2012) examine the mechanisms behind knowledge spillovers of suppliers. While the authors don't find evidence for a supporting role of suppliers' absorptive capacities on knowledge acquisition, they find evidence for a supportive role on performance improvement and new knowledge creation.

These studies, however, do not examine the underlying transmission channel through which positive FDI spillovers on suppliers occur. New research by Godart and Görg (2013) sheds light on this issue. The authors find evidence for a "forced linkage effect" (or demand effect) from supplying multinationals. Suppliers in a cross-section of 25 Eastern European and Central Asian countries experienced productivity growth that was related to "pressure from their customers." Interestingly, productivity spillovers cannot be observed in the absence of this "forced linkage effect," nor for more cooperative learning mechanisms such as technology transfer by the multinational.

Another recent study uses the U.K. Workplace Employment Relations Survey across a sample of manufacturing and services industries and finds evidence for horizontal and vertical spillovers of management practices from multinationals to local firms (Fu 2012). Backward and forward spillovers of management practices within supply chains, in particular, are found to be economically larger than horizontal spillovers, implying that supply chains play an important role as transmission channel.

Spillovers through Labor Turnover

Investment of multinationals in their workforce provides workers with knowledge and skills, the benefits of which may not be completely internalized, as knowledge may be carried over to local firms—horizontally or vertically—through labor turnover. Thus, knowledge embodied in the labor force may move from multinational to local firms—either to existing local firms or when workers start their own firms (Fosfuri, Motta, and Ronde 2001; Glass and Saggi 2002; Crespo and Fontoura 2007). For example, Quddus and Rashid (2000; cited in Saggi 2002) describes how 130 workers of a Bangladeshi domestic apparel firm that supplied a Korean firm were sent for training to the Republic of Korea: 115 of these workers returned to set up their own apparel firms or joined other newly established domestic firms. Similarly, Rodrik (2003) argues that entrepreneurial entry through labor turnover drove the huge growth of Chile's salmon industry in the 1980s and 1990s.

However, spillovers through labor are more likely to be realized in the medium to long term, as knowledge first needs to be absorbed by the local workforce. In the short term, it is more likely that foreign firms will bid away high-quality labor from domestic firms by offering higher wages and benefits, resulting in a potentially negative spillover effect (Sinani and Meyer 2004; Hoekman and Javorcik 2006; Crespo and Fontoura 2007). Aitken, Harrison, and Lipsey (1996), for example, find for a sample of manufacturing plants in Mexico and República Bolivariana de Venezuela that FDI results in lower wages in domestic firms; the authors relate this to foreign firms poaching skilled workers from domestic firms. Loss of skilled labor also affects the domestic firm's capacity to benefit from FDI spillovers.

Recent studies evaluated the existence of labor turnover as a channel for productivity spillovers from FDI. Balsvik (2011), for instance, finds that employees in Norwegian manufacturing firms with previous work experience in multinationals contribute 20 percent more to productivity than their counterparts without such an experience. Interestingly, the productivity premium of workers with previous experience in multinational firms seems to be driven by less-skilled workers (with less than 12 years of schooling), while the share of skilled workers that previously worked in a multinational firm has no significant effect on productivity. This confirms the results of Görg and Strobl (2005), who find that for a sample of manufacturing firms in Ghana, domestic entrepreneurs with previous work experience in a multinational firm in the same industry run more productive firms than entrepreneurs without such an experience. However, the results also suggest that such firm owners benefit more strongly if they have a lower educational level. The explanation given is that the entrepreneurs' capacity to absorb new knowledge is larger for individuals with a lower educational level, because of the steeper learning curve they face.

Spillovers through labor turnover depend on the ability of domestic firms to attract labor that previously worked in multinational firms. Competitive wages are likely to be the most important driver to attract the local workforce which, in turn, may be a function of firm size, a higher technology and R&D intensity, and openness to trade. Spillovers also depends on the ease of and support for establishing new local firms, which is affected at the host country level by investment policies, access to finance, and small and medium enterprise (SME) development policies. The extent of spillovers also depends on the amount of training and knowledge local workers receive in multinational firms; the share of local workers versus expatriates in skill-intensive technical, supervisor, or management positions—which can also be influenced by the degree of foreign ownership and the global production strategy of the investor; and the proximity to and interactions with local firms. At the host country level, the extent of labor market rigidity determines whether labor can flow freely between firms, but also how long workers stay with a firm, an important factor for knowledge absorption and diffusion.

Spillovers through Changing Market Forces—Competition, Demonstration, and Collaboration

Competition Effect

Foreign entry can result in increased competition in product, labor, and credit markets between local and foreign producers, especially if the foreign firm also sells to the local market. This competition effect refers to "pecuniary externalities" (Javorcik 2008). In the short to medium term, local firms might face losses in their output and market share. This, in turn, requires them to produce at higher average unit costs due to declining economies of scale (Harrison 1994; Aitken and Harrison 1999; Crespo and Fontoura 2007). In the long run, the average performance, including productivity, quality, and reliability, of local producers might improve due to the necessity to keep up with foreign competitors and the exiting of the worst performers. This, in turn, will benefit downstream sectors (Javorcik 2008). The presence of multinational firms may also increase competition among local firms that want to become their suppliers, resulting in a higher quality and reliability of inputs (Crespo and Fontoura 2007; Javorcik 2008). Higher competition in supplying sectors might go hand-in-hand with complementary domestic investment in these sectors (Kugler 2006).

Certain domestic firm characteristics might influence their capacity to bear up against stronger competition and eventually even benefit from it.

These include, most importantly, a domestic firm's productivity and technology gap relative to the foreign competitor, R&D capacity, worker skills, firm size, and participation in export markets. Aspects of the host country's policy environment may also contribute to a domestic firm's capacity to compete and internalize FDI spillovers, including the degree of existing market competition; access to finance; learning and innovation infrastructure; and trade, investment, and industrial policies. Characteristics of the foreign investor also play a role in determining the degree of competition, including its technology intensity, location, FDI motive, entry mode, and its global production and sourcing strategy.

Demonstration Effect

As local producers are exposed to multinational firms' products, marketing strategies, and production processes, knowledge spillovers from direct imitation or reverse-engineering might accrue. Referred to as the "demonstration effect," this type of knowledge diffusion represents "real externalities" from foreign entry (Javorcik 2008). Observing foreign firms successfully using advanced technologies, processes, and techniques might encourage local producers to incorporate these tools as well, as the demonstration effect reduces their uncertainty about the effectiveness of such tools (Crespo and Fontoura 2007). Demonstration effects are most likely to benefit local producers within the same sector (horizontal spillovers), but vertical demonstration effects may also confer advantages to local suppliers or buyers, to be discussed further.

Demonstration effects also include foreign firms' know-how about export processes with respect to distribution networks, transport infrastructures, and consumers' tastes in destination markets, among other things (Greenaway, Sousa, and Wakelin 2004). Local firms can reduce entry costs to exporting by imitating foreign firms' export processes or collaborating with the latter (Crespo and Fontoura 2007). This mechanism is particularly beneficial for local suppliers who seek to become exporters. Additionally, the host country may emerge for the first time on the exporting and sourcing landscape, because of the entry of foreign firms with established relationships to global buyers or manufacturers. If these global companies get to know the country and see it as a main supplier, then it may also be easier for local firms to access these networks.

Again, several factors mediate the impact of FDI on local firms via the demonstration effect. Domestic firms are more able to imitate foreign firms if the latter dispense a certain amount of know-how in the form of, for example, technology, R&D, human capital, or export knowledge. Domestic firms' willingness to restructure and imitate also depends on the type of ownership and seems to be higher for private firms. Certain mediating factors in the host country can further influence the domestic firms' capacity to imitate, such as the learning and innovation infrastructure. Finally, foreign firm characteristics influence the type of FDI entering the country. The spillover potential via the demonstration effect, for instance, strongly depends on the foreign firm's technology intensity, FDI motive, its proximity to and interactions with local firms, and the extent to which foreign knowledge can leak. Host country characteristics and institutional framework, such as intellectual property rights and governance, in turn, influence the type of foreign investors being attracted and the knowledge they incorporate.

The Role of Mediating Factors

Foreign Firm Characteristics

Degree and Structure of Foreign Ownership

The degree of foreign ownership affects local firms' potential to absorb FDI spillovers, although the direction of this effect is ambiguous. On the one hand, a higher share of foreign ownership, and, thus, greater control over management, correlates positively with the parent firm's incentive to transfer knowledge, for example, in the form of technology. This has been confirmed by empirical studies for Greece (Dimelis and Louri 2002) and Indonesia (Takii 2005). On the other hand, a larger domestic ownership share could also be beneficial for local firms, since the foreign investor's interest in reducing spillovers might be less well protected, making technology leakages more likely (demonstration effect). Greater domestic participation has also been shown to increase linkages with domestic suppliers given better information on and relations to potential suppliers (see, for example, Toth and Semjen 1999). The labor market spillover channel may also increase in importance in joint ventures (JVs), as locals are involved at the owner and top management level, which not only contributes directly to skill development but also increases the tendency of using locals in higher positions.

Empirical studies controlling for different structures of foreign ownership tend to support the more positive spillover effects of JVs. For example, Havranek and Irsova (2011) find evidence for lower spillovers in fully owned foreign affiliates, and Javorcik (2004a) and Javorcik and Spatareanu (2008) find a positive vertical spillover effect on domestic firms in supplying industries from multinationals with partial foreign ownership, but not from multinationals with full foreign ownership. Abraham, Konings, and Slootmaekers (2010) find for a sample of Chinese manufacturing firms that foreign ownership in a domestic firm's sector only results in positive horizontal spillovers when foreign ownership is organized as a JV. By contrast, the presence of fully owned foreign firms is found to have a negative impact on local firms, due to the technology intensity of multinationals crowding out local producers within the same sectors (Abraham, Konings, and Slootmaekers 2010).

Length of Foreign Presence

In addition, the length of foreign presence of a multinational in the host country also moderates FDI spillovers. Gorodnichenko, Svejnar, and Terrell (2007), for example, focus on FDI spillovers from old versus new firms in 17 Central and Eastern European transition economies, Turkey, and the Commonwealth of Independent States; they find significantly positive forward and horizontal FDI spillovers from older firms (established before 1991), while these effects cannot be confirmed for newer firms (established in or after 1991). An explanation may be that older firms know the local context better and hence available local suppliers and labor skills.

FDI Motive

It is widely accepted that different motivations for undertaking FDI—including resource seeking, efficiency seeking, market seeking, and asset seeking—are likely to mediate spillover potential. However, evidence remains scant that one form of FDI is inherently more or less prone to spillovers. The conventional wisdom on resource-seeking FDI is that it has limited potential for spillovers, due to its high capital and technology intensity, limited time horizons, and enclave character. Evidence from Chile suggests that while FDI in natural resources tends to use "state-of-the art" technologies within their sectors, these technologies seem to be "undemanding" compared to other sectors. Moreover, there is limited evidence for spillovers from training measures of foreign firms due to the capital-intensive nature of such investments (Pigato 2001). On the other hand, recent research[2] has identified that, in particular, backward linkage potential exists and is growing, both in terms of material inputs as well as services (Morris, Staritz, and Barnes 2011). This is driven in part by increasing trends of large resource-extracting firms to outsource the production of inputs that are outside of their core competences, as well as increasing pressures by some host governments for local sourcing.

By contrast, it is often considered that FDI in the manufacturing sector has higher spillover potential as it is largely driven by efficiency-seeking motives. Indeed, the more labor-intensive nature of manufacturing investment, its requirements for a broad range of goods and services inputs, and the lower barriers to domestic forward linkages (relative to resource-seeking FDI) make it a strong candidate for contributing spillovers. On the other hand, much of the "export assembly" activity that characterizes labor-intensive global production network investment operates with similar motivations and strategies as resource-seeking FDI—essentially "extracting" low-wage labor rather than broader capabilities of local firms. Therefore, the spillover potential of efficiency-seeking FDI varies most importantly with the technology and skill intensity of the production process, and with the production and sourcing networks and GVC dynamics in different manufacturing sectors.

Market-seeking FDI, in particular in retail, is also thought to provide higher spillover potential as retailers tend to source from local producers, in particular for food and other perishable products. Lesher and Miroudot (2008), for example, find stronger backward spillovers on domestic firm productivity from FDI in services sectors across a sample of 15 Organisation for Economic Co-operation and Development countries located in Europe. However, this effect tends to be sector and retailer specific, and importantly depends on the capabilities of local suppliers and the support of or requirements for local sourcing by host governments. Barrientos and Visser (2012) confirm the findings of other studies on horticulture sectors in Sub-Saharan Africa that larger plantations can more easily comply with standards of supermarkets than small- and medium-scale producers. However, in the context of contract farming becoming more important, smaller firms may have access to supermarket networks through larger firms that organize a network of local contractors.

While asset-seeking FDI also refers to a multinational's motive to extract technology- or skill-intensive "resources" from a country, the FDI spillover potential may be higher due to closer relationships with local suppliers, customers, and workers. Giroud, Jindra, and Marek (2012), for example, find evidence that a multinational's "external technological embeddedness"—that is, the importance of collaboration in R&D with local suppliers or customers for the multinational's own R&D or innovation—has a positive effect on the intensity of backward linkages in a sample of five transition countries.

Global Production and Sourcing Strategy
The possibility for local firms to supply a multinational depends importantly on the multinational's global production strategy. If production is highly internalized, for example, because a large share of value added is considered a core competency, then the multinational is likely to have little interest in local sourcing beyond nontradable services and standardized inputs like packaging materials (Paus and Gallagher 2008). By contrast, if lead firms concentrate on high-value activities such as marketing, branding, and design, then there is more scope for domestic firms to take over other activities. In manufacturing sectors like automotive, apparel, footwear, and electronics, there has been a strong trend toward externalized production models—with differences, however, related to FDI home countries. For example, in the electronics sector in Malaysia, Japanese firms tended to pursue a more internalized production model whereas U.S. investors were more open to outsourcing to local firms (Phillips and Henderson 2009; see also discussion that follows). While this outsourcing strategy began in manufacturing operations, it is also seen increasingly in natural resource sectors, where large resource-extracting firms have begun outsourcing activities outside of their core competences (Morris, Kaplinsky, and Kaplan 2011).

Analogously, a multinational firm's global sourcing strategy may affect the spillover potential. If a multinational firm sources on a global scale, it may follow a co-sourcing strategy that relies on imported inputs from established suppliers abroad. Co-sourcing is particularly relevant for multinationals that have production plants in more countries. It ensures that the same inputs are used in their global production with regard to type, quality, color, and specifications. Co-sourcing also obtains better prices because inputs are sourced in larger quantities (for the apparel sector, see Staritz and Morris 2012, 2013). Alternatively, a multinational firm might follow co-location strategies that require established foreign input suppliers to co-locate with them and hence also enter the host country. Both co-sourcing and co-location render the entrance of new local suppliers more difficult (Paus and Gallagher 2008), but in the latter case local investment and employment is extended, opening up at least some potential for spillovers at the second-tier level. Giroud, Jindra, and Marek (2012), for example, find for a sample of five transition countries that a foreign affiliate's greater autonomy in basic and applied research activities increases the intensity of linkages with suppliers in the host country. However, the study cannot confirm the effect for a greater autonomy in production and operational management decisions.

However, sourcing policies are not only determined directly by foreign investors but also by lead firms or the buyers of the foreign investor, including retailers or brands. In the apparel sector, for example, it is common that global buyers such as retailers (such as Wal-Mart, The Gap, Marks and Spencer, and H&M) or brands (such as Levi and Nike) nominate suppliers of main inputs such as textile and trims that are needed by their suppliers producing apparel (Staritz 2011). This is also common in automotive and electronics (Plank and Staritz 2013).

More recently, and with variations at the sector level, there has been a trend toward supply chain consolidation strategies (Cattaneo, Gereffi, and Staritz 2011). This results in outsourcing being carried out more often through a supplier tiering model. For example, all production-related (and, increasingly, nonproduction-related) operations are passed on to a large, "first-tier" supplier, which then manages its own production or sourcing network. It is typically difficult for domestic firms to break in as first-tier suppliers, particularly in sectors where inputs are technologically complex (for example, automotive and electronics as well as oil, gas, and mining), as lead firms typically have long-established, accredited global supplier networks. In such situations, the co-location of foreign first-tier suppliers may at least provide opportunities for local firms to enter as second-tier suppliers, although many first-tier suppliers follow a more internalized production model.

Another factor relates to technology, scale, and diversity requirements. If the production of inputs requires a high degree of technological sophistication, then a multinational firm may opt to purchase inputs from existing suppliers with which it already has long-standing relationships (Paus and Gallagher 2008). The size and diversity of inputs needed may also affect local sourcing decisions. If multinational firms cannot source the full volume of inputs locally, then they may opt to import from countries that can offer a "one stop shop," meeting volume and diversity requirements. In the apparel sector, for example, some locations in China offer a full range of textiles and trims in large quantities, making it possible for apparel manufacturers to source all required inputs from a single location. This trend significantly limits the potential for domestic suppliers.

Technology Intensity

FDI spillovers depend on the technology intensity of the goods and services produced by the multinational in the host country. More products that are more technology (or R&D) intensive generally contain a greater element of knowledge and broader set of skills. Focusing on FDI in technology-intensive industries, Buckley, Wang, and Clegg (2007) find positive spillovers on Chinese firms to be stronger if originated by Western-owned multinationals compared to affiliates from Taiwan, China; Hong Kong SAR, China; and Macao SAR, China. The authors relate this to the higher technology intensity in Western-owned multinationals. Analogously, Lin, Liub, and Zhanga (2009) differentiate between FDI in China from Taiwan, China; Hong Kong SAR, China; Macao SAR, China; and FDI from other countries. They confirm the positive horizontal and vertical spillovers for FDI from other countries. However, FDI from Taiwan, China; Hong Kong SAR, China; and Macao SAR, China results in positive forward spillovers, but no backward spillovers, and negative horizontal spillovers. This result is also explained by the more labor-intensive nature of foreign affiliates from these economies (Lin, Liub, and Zhanga 2009).[3]

Using a direct measure of the foreign investors' technology intensity, Giroud, Jindra, and Marek (2012) confirm that the multinationals' technological capability, measured in terms of their relative innovation intensity, has a significantly positive impact on the intensity of backward linkages in a sample of five transition countries. High-tech FDI may also lead to limited spillovers, if the technology gap between foreign and local firms is so large that domestic firms are unable to absorb potential spillovers (see the next subsection on domestic firm characteristics).

In this context, it is important to make a distinction between the technology content of the sector or end product and the technology content of the specific processes and outputs carried out in the host country. In GVC-organized production, it is often the case that knowledge- and capital-intensive (high-tech) processes are separated geographically from the labor-intensive (low-tech) processes of the same product (Plank and Staritz 2013). Production in developing countries is, therefore, more likely to focus on labor-intensive, low- to medium-tech activities and rely on high-tech imported inputs from other countries. Although this reduces the FDI spillover potential through supply links, such high-tech imports may lead to "import spillovers" in so far as imports contain technology and knowledge that can be absorbed by domestic markets.

FDI Home Country

As previously discussed, the FDI home country may have an effect on the production strategy pursued and on the technologies used in host countries. It may also have other effects on spillover potential. The home country of FDI influences managerial practices and cultures, which are related to differences in the use of expatriate workers, attitudes and strategies concerning the training of local workers, and general skills development. Further, end-market segmentation—closely linked to FDI home countries through historical, cultural, and language ties, as well as trade policies—is a common practice (Cattaneo, Gereffi, and Staritz 2011). In the apparel sector, for example, European-owned firms in Mauritius and Madagascar largely export to Europe, whereas Asian-owned firms serve the U.S. market; in Lesotho, Asian-owned firms export to the United States while South African–owned firms export to South Africa (Gibbon 2003; Staritz and Morris 2012, 2013). These patterns impact on spillover potential, as buyer sourcing requirements and practices can vary considerably by market. Moreover, production for one specific market may bring a firm setup and overhead structure that is uncompetitive for other markets (Gibbon 2003).

Zhang et al. (2010, 970) find that for a panel of Chinese manufacturing firms between 1998 and 2003, a domestic firm's exposure to "different systems of technologies, management practices, and cultural values brought by foreign firms from different country origins" increases the possibility for positive FDI spillovers. The study gives two major explanations for this effect. On the one hand, a greater variety of foreign firm practices can positively influence the domestic firm's absorptive capacity, for example, by increasing its openness to learning and exploring new knowledge. In addition, a greater variety of foreign firm practices can enhance the impact of the transmission channels, as it creates more opportunities to learn from observation and imitation and a greater variety of knowledge transmission via supply chains and labor turnover (Zhang et al. 2010).

Finally, a foreign affiliate's distance to its parent firm impacts its spillover potential, particularly for efficiency-seeking FDI. Several studies find that foreign investors are more likely to purchase local inputs from domestic suppliers if the home country is farther away (Havranek and Irsova 2011; Javorcik and Spatareanu 2011). The underlying reasoning is that the cost of communications and coordination increases with distance (Rodriguez-Clare 1996). However, the role of geographic distance varies by sector; for example, distance clearly impacts services more than goods. Thus, a multinational firm's global production and sourcing strategy may well become a more important determinant of FDI spillover potential than geography alone.

Entry Mode

A multinational firm's entry mode may influence the extent or pace of FDI-induced benefits for local firms. For example, a greenfield investment is more likely to be accompanied by the implementation of leading technology. By contrast, in the case of a merger and acquisition (M&A), the multinational firm is more likely to adopt the host country's technology, and to only gradually improve its technology (Crespo and Fontoura 2007). Branstetter (2006) confirms higher spillover effects from Japanese greenfield affiliates in the United States as opposed to total investments, but rejects the existence of spillovers from acquisitions. Braconier, Ekholm, and Knarvik (2001) also argue that integration of foreign knowledge in M&As tends to be a gradual process, thus making leakages to other domestic firms less likely, whereas knowledge transfer in greenfield investments happens more quickly. Additionally, while greenfield investments by definition increase investment, capacity, and employment, M&A and other types of brownfield investments may not, as the new foreign owners may rationalize and even reduce capacity and employment.

Another recent study finds that the pace and irregularity of foreign entry matters for the realization of FDI spillovers;[4] specifically, quick and irregular foreign-entry processes were found to reduce FDI benefits for Chinese manufacturing firms (Wang et al. 2012). This may be because the speed and irregularity of foreign entry prevent multinationals from building stable relationships with local suppliers, making it less likely that foreign firms will rely on domestic inputs. Furthermore, local firms and workers might not have enough time to observe and imitate good practices and to acquire skills, resulting in negative competition effects.

Domestic Firm Characteristics
Technology and Productivity Gap

The technology gap between foreign and domestic firms has been identified as one of the most important factors mediating FDI spillovers (Kokko 1994; Kokko, Tansini, and Zejan 1996; Grünfeld 2006). The idea of a "technology gap" is that the efficiency and modernity of equipment and production processes employed by local and foreign firms may differ considerably. It is usually measured as a domestic firm's productivity level relative to a benchmark productivity level within the same sector—often of the leading firms (Griffith, Redding, and Simpson 2002; Girma 2005; Girma and Görg 2007) or of foreign firms (Castellani and Zanfei 2003).

Views on the role of the technology gap for FDI spillovers conflict. On the one hand, some studies find that a large technology gap is beneficial for local firms, since their catching-up potential (and, therefore, the potential for spillover from FDI) increases (Findlay 1978; Wang and Blomström 1992; Smeets 2008; Jordaan 2011). Jordaan (2011), for instance, finds that a local firm's experience it has gained from supplying inputs to local producers positively affects FDI spillovers. Other studies argue that local firms are less able to absorb spillovers if the technology gap between the multinational and local producers is too big. Spillovers, in particular with regard to linkages, depend on the capacity of local firms to satisfy the demands of multinational firms at satisfactory cost and quality levels.

Girma (2005) reconciles both views and finds a nonlinear relationship between a domestic firm's technology gap and FDI-induced productivity benefits for a sample of U.K. manufacturing firms. That is, above and below a certain threshold gap, spillover effects are limited or even negative; only between these thresholds do reductions in the technology gap increase productivity gains from FDI (Girma 2005). Girma and Görg (2007) confirm a similar U-shaped relationship for domestic firms in the U.K. electronics and engineering sectors. Blalock and Gertler (2009) also reject the existence of spillovers for Indonesian manufacturing firms with a small technology gap.

Research and Development

While productivity and technology are closely related,[5] another factor that may be considered in assessing absorptive capacity is the level and role of R&D in local firms. The literature suggests that there is solid evidence for the supportive role of R&D in local firms for developed countries. Examples include Spain (Barrios and Strobl 2002; Barrios et al. 2004), the United States (Keller and Yeaple 2009), Ireland (Barrios et al. 2004), and Sweden (Karpaty and Lundberg 2004), among others. There are also studies confirming the supportive role of R&D in domestic firms for developing or emerging countries, including the Czech Republic (Kinoshita 2001), India (Kanturia 2000, 2001, 2002), Hungary and the Slovak

Republic (Damijan et al. 2003), and Indonesia (Blalock and Gertler 2009). One exception in the literature is Damijan et al. (2003), who find a negative role of firm-level R&D on FDI spillovers for Estonia and Latvia (reported in Crespo and Fontoura 2007). Studying a sample of five transition countries, Gentile-Lüdecke and Giroud (2012) find no impact of suppliers' R&D intensity on their knowledge acquisition from multinationals, but there is an impact on local suppliers' new knowledge creation in terms of new products, services, and technologies.

Human Capital
A domestic firm's ability to absorb foreign technology might also be positively related to its share of skilled labor. Blalock and Gertler (2009), for example, find that the proportion of employees with college degrees significantly increases domestic firms' productivity gains from FDI in the Indonesian manufacturing sector. However, Girma and Wakelin (2007) only confirm such a finding for small firms in the United Kingdom. They find that FDI does not affect large firms with a high proportion of human capital, as these firms are probably the most similar to multinationals in terms of technology and market share.

In contrast, Sinani and Meyer (2004) find that, for a sample of Estonian firms, a larger share of human capital reduces the positive spillover effects for small firms, but increases it for large firms. Their explanation for this contradicting result is that the competition effect might reduce workers' possibility of extracting additional rents from local firms, since multinationals tend to pay better wages. The competition effect might also enable larger firms to keep skilled workers, compared to smaller firms that might lose skilled workers to foreign firms.

Scale
Firm size has been positively related to a domestic firm's capacity to absorb FDI spillovers (see, for example, Jordaan 2011), perhaps because firm size tends to be positively correlated with productivity. Larger firms may be better positioned to compete with multinationals and to imitate their tools (Crespo and Fontoura 2007), as well as to fulfill supply requirements and standards and provide the volumes demanded by multinationals. Analogously, larger firms may pay better wages and therefore find it easier to attract workers employed by multinational firms. Larger firms might also be more visible (for example, through membership in associations), and thus more likely selected as local suppliers by foreign firms. While Aitken and Harrison (1999) find negative spillovers from FDI for domestic plants in República Bolivariana de Venezuela, these effects are only significant for firms with fewer than 50 employees. This suggests that smaller firms may be less productive and less capable of absorbing positive spillover effects. In contrast, other studies find that SMEs benefit more strongly from FDI spillovers, especially those firms with a higher proportion of skilled labor (see, for example, Girma and Wakelin 2007; Sinani and Meyer 2004). In contrast, Gentile-Lüdecke and Giroud (2012) find evidence for a negative effect of firm size on knowledge acquisition from multinationals for suppliers in the Polish automotive sector.

Besides firm size, sector size is also important, as it ensures a scale of possible suppliers that can offer volumes required by multinational firms. Finally, the size of a region or country imposes limitations on the number of fields in which the country can have absorption capacity. Limited human and physical resources necessitate more specialization, particularly in fields where economies of scale do not play a substantial role (Paus and Gallagher 2008).

Firm Location

Several aspects of domestic firm location have been shown to be important for the extent of spillovers from FDI. First, agglomeration economies—that is the co-location of foreign and domestic firms in the same region—can mediate the benefits from FDI on domestic firms. Barrios, Bertellini, and Strobl (2006) find evidence that foreign firms co-locating in the same sector and region significantly increase productivity and employment of local manufacturing firms in Ireland. Some studies contest the positive role of co-location for a firm's absorptive capacity. While Sjöholm (1999) confirms positive spillover effects when FDI is measured at the country-sector level in Indonesia, he finds negative spillovers when foreign presence is measured at the region-sector level. Aitken and Harrison (1999) find a similar result for República Bolivariana de Venezuela and Yudaeva et al. (2003) for the Russian Federation. Closely related, co-location of firms in industry clusters has been shown theoretically and empirically (Nadvi and Schmitz 1994; Thompson 2002) to have an important impact on spillovers.

Second, regional characteristics also play an important role in mediating FDI spillovers. Firms located in regions with better access to skilled labor or research facilities, for instance, may have a higher absorptive capacity compared to firms located in lagging regions. The significant role of firm location within regions that are more advanced, for example, in terms of education or R&D intensity, is confirmed in the case of Russia (Ponomareva 2000), the United Kingdom (Girma 2005), and the Czech Republic (Torlak 2004), among others. Girma and Wakelin (2007) also find that positive FDI-induced horizontal productivity spillovers are less pronounced for domestic electronics firms located in lagging regions (defined as regions with assisted area status) within the United Kingdom.

Third, space—that is a domestic firm's geographical distance from multinational firms—can affect the extent of FDI spillovers. Girma and Wakelin (2007), for instance, show evidence that spillovers seem to be constrained to the region where the multinational is located. Resmini and Nicolini (2007) confirm positive productivity gains for domestic firms from foreign firms within the same region, but find negative spillovers from foreign firms located outside the region. Export processing zones with a large share of foreign firms and limited access for local firms, due to regulatory barriers or requirements for large investments, may constrain spillovers.

Exporting

Exporting has been linked to a domestic firm's absorptive capacity for at least two reasons. First, local exporting firms are generally characterized by a higher productivity—be it via learning-by-exporting or self-selection into exporting—rendering them more competitive against negative rivalry effects created by multinationals (Crespo and Fontoura 2007). Firms with a track record as exporters may also be in a better position to be nominated as an input supplier or subcontractor to a multinational. Second, the more a local firm exports, the lower will be competitive pressures from multinational firms (assuming that the multinational firm doesn't enter the same export market), thus reducing incentives to improve, which lowers the extent of positive FDI spillovers. In fact, several empirical studies confirm that the potential for positive productivity spillovers from FDI is less pronounced for exporters compared to nonexporters or firms exporting little; these studies include Blomström and Sjöholm (1999) for Indonesia, Ponomareva (2000) for Russia, Sinani and Meyer (2004) for Estonia, and Abraham, Konings, and Slootmaekers (2010) and Du, Harrison, and Jefferson (2011) for China.

In contrast, some studies find that the gains from FDI are larger for exporting firms, for example, Jordaan (2011) for domestic suppliers in Mexico. Similarly, other studies find more positive intrasectoral spillovers for local firms operating in more open sectors, for example, Barrios and Strobl (2002) for Spain, and Schoors and van der Tol (2002) for Hungary (the latter also finding more positive backward spillovers). Although Lin, Liub, and Zhanga (2009) reject that export intensity has an effect on horizontal FDI spillovers on domestic firms in China, they confirm larger positive backward and forward spillovers. However, the additional spillover benefits from export intensity become smaller the larger is a firm's export intensity (that is, the greater share of its output that is exported). These contradicting results may be explained by a positive demonstration effect. Firms exposed to stronger competition might be better prepared to imitate good practices from multinational firms.

Sector Dynamics

Spillovers also depend on the sectors in which domestic firms operate. This has been discussed extensively in the GVC literature (see, for example, Cattaneo, Gereffi, and Staritz 2011). Temenggung (2007) studies the effect of horizontal FDI spillovers for several manufacturing sectors in Indonesia over the period 1975–2000 and finds a positive effect only in "food and beverages" and "textiles and leather." Suyanto and Salim (2010) go one step further and detect the nature of the spillovers contributing to FDI-enhanced productivity growth in two sectors in Indonesia between 1988 and 1995. They conclude that FDI-enhanced productivity spillovers in the food-processing industry seems to be driven by efficiency improvements, while technological progress seems to be the main driver in the electrical machinery industry.

Tondl and Fornero (2010) use sectoral data to study the effect of FDI spillovers in 14 Latin American countries for eight broad sectors over the period 1990–2006. They find that both the agriculture & fishing and mining sectors show significantly positive productivity effects from intrasectoral FDI as well as intersectoral FDI. Havranek and Irsova (2011) find that spillovers are smaller for domestic firms in services sectors due to lower absorptive capacities of firms in these sectors.

Other studies specifically compare the extent of spillovers in technology-intensive industries to other industries. Buckley, Wang, and Clegg (2007) find, for instance, that foreign presence in technology-intensive industries in China led to larger positive spillovers on Chinese firms' output in comparison with labor-intensive industries. Keller and Yeaple (2009) find that technology spillovers from FDI to the United States between 1987 and 1996 were substantially larger for domestic firms in high-tech sectors than in low-tech sectors. This is possibly related to the technology intensity of foreign investors, as discussed in the previous subsection on foreign firm characteristics.

Type of Ownership

Another factor determining a domestic firm's absorptive capacity is the type of ownership. Some studies have focused on the difference between private versus state-owned firms, which can be studied best in the context of China or the transition economies in Central and Eastern Europe. Private firms may be more likely to benefit from FDI spillovers due to their willingness to restructure and imitate (demonstration effect) and a larger export orientation, enabling these firms to access knowledge internationally (Sinani and Meyer 2004). On the other hand, state-owned enterprises (SOEs) are typically larger, technically competitive, and may have easier access to finance, increasing their absorptive capacity. However, they tend to be less market-oriented (Du, Harrison, and Jefferson 2011), which may lower absorptive capacity.

Li, Liu, and Parker (2001) find that private and collectively owned firms in China benefit from demonstration effects, whereas SOEs benefit from competition effects. Focusing on a sample of Estonian firms, Sinani and Meyer (2004) find that the positive productivity gains from FDI are larger for "outsider-owned" private firms than for either "insider-owned" private firms (that is, owned by management and/or staff) or SOEs. One explanation given is that insider-owned firms are smaller, more labor-intensive, and have less access to finance, which reduces their absorptive capacities. Outsider-owned firms, in addition, are more successful in improving performance due to privatization, and hence are also expected to better cope with foreign presence (Sinani and Meyer 2004). In contrast, Lin, Liub, and Zhanga (2009) find that both state-owned and non-state-owned firms benefit from backward and forward FDI spillovers, but these tend to be higher for SOEs.

Competition

Finally, the level of competition also influences the extent of FDI spillovers. Competitive pressures from multinational firms might be lower if the local firm already faces a high level of competition at the sectoral level. As in the case of exports, local firms in competitive sectors might have a lower incentive to improve, resulting in lower benefits from FDI spillovers. On the other hand, local firms could be better equipped to benefit from positive demonstration effects.

Using income per capita as a proxy for national competition, Meyer and Sinani (2009) combine both views. The authors provide evidence for a U-shaped relationship wherein the extent of FDI spillovers is larger for countries below the minimum threshold level ("no competition") and above the maximum threshold level ("dynamic competition") of development. Between these threshold levels, FDI spillovers are smaller because of a "crowding-out" effect on local firms (Meyer and Sinani 2009).

Host Country Factors and the Institutional Framework

Host country factors and the specific institutional context act as mediating factors for FDI spillovers in a threefold manner, as depicted in figure 2.1. First, they influence foreign investors' characteristics, and thus the spillover potential. Second, they shape the characteristics of domestic firms and, thus, their absorptive capacity. Third, they affect the transmission channels from foreign to domestic firms. There may be cases where a foreign investor has a high FDI spillover potential (for example, new technologies) while a domestic producer has a high level of absorptive capacity (for example, a high share of skilled labor), but despite these conditions, domestic firms cannot (fully) internalize positive spillovers due to factors impeding the transmission channels. Alternatively, there may be factors within these transmission channels that facilitate greater spillovers. Labor market regulations, for instance, can affect the mechanism of labor turnover in a country, while the regulatory, market, and governance environment will influence the functioning of the competition and demonstration effects. The regulatory context and trade, investment, and industrial policies will influence the extent of supply chain spillovers. In this section, we examine various host country policy and institutional factors that may mediate the direction and extent of FDI spillovers.

Labor Market Regulations

Labor market regulations can influence the effect of FDI on domestic firms via various channels, including the amount and type of FDI being attracted in the first place, the potential for knowledge absorption, and the transmission of these spillovers into the domestic economy. Javorcik and Spatareanu (2005), for instance, find that higher absolute and relative labor market flexibility compared with the foreign investor's home country had a positive impact on the likelihood of foreign investment. Whether this translates into spillovers depends on the absorptive capacity of domestic firms. Evidence suggests that labor market regulations affect this, too, particularly through worker skills. In a recent study, Hale and Long (2011) find that FDI in China puts upward pressure on wages of skilled workers. While private firms manage to compete with foreign firms for skilled workers by paying higher wages (competition effect), SOEs—which are more constrained in terms of the wages they can pay—have to accept skilled labor of lower quality. The authors conclude that labor market regulations in general, and wage constraints in particular, affect the level of skills in a firm and, thus, its absorptive capacity.

In addition, labor market regulations affect the domestic firm's willingness to invest in job training, which indirectly influences skill availability. Using firm-level data across 64 developing countries, Almeida and Aterido (2011) find, for example, that rigidity of employment increases a firm's willingness to invest in job training. Examining the type of labor market regulations in more detail, the authors find that this positive effect holds only for difficulty of hiring and rigidity of hours, while the difficulty of firing has no impact. In sum, labor market regulations seem to affect firm training and, in turn, worker skills and domestic firms' capacity to absorb FDI spillovers.

Labor market regulation can also be strategically used in the form of increasing wages and improving working conditions to force firms to become more productive and develop into higher-value activities. This has been used in some East Asian countries in the context of other industrial policy measures. Higher wages and better working conditions can also contribute to increased productivity, given the improved circumstances for workers or their higher motivation (for literature surveys on the efficiency wage see, for example, Yellen 1984; Katz 1986; Weiss 1991).

Finally, labor market regulations directly affect the frequency and nature of labor turnover and, thus, the transmission channel. FDI spillovers through labor turnover in a highly rigid labor market context may be limited. On the other hand, too flexible labor markets may result in high churning of workers, lowering the incentive for firms to invest in training and the possibility for domestic workers to acquire skills and knowledge, but also lowering incentives for acquired knowledge to diffuse to local firms.

Intellectual Property Rights

The strength of intellectual property rights in a host country have an impact on the quality of foreign investment that can be attracted, and therefore the potential for FDI spillovers (Gorodnichenko, Svejnar, and Terrell 2007). Javorcik (2004b), for example, finds that for a sample of firms in Central and Eastern European countries, the likelihood of attracting foreign investors in sectors that rely on strong intellectual property rights (for example, high-tech products) is lower for host countries with weak property rights protection. Branstetter, Fisman, and Foley (2006) confirm that U.S. multinational firms transferred more technology to countries that improved their intellectual property rights between 1982 and 1999. Foreign investors are also more likely to distribute imported products instead of producing locally when intellectual property rights are weak (Javorcik 2004b).

Havranek and Irsova (2011), however, reject the hypothesis that a country's level of protection of intellectual property rights influences the magnitude of FDI spillovers on domestic firms. Strong intellectual property rights may help attract FDI with higher spillover potential (as more knowledge is transferred to the affiliate). However, the same strong rights may also act as a barrier to the transmission of knowledge and technology to the local market. Multinational firms use several instruments in addition to strong property rights to protect technology spillovers to local competitors in the same sector, such as paying higher wages to avoid labor turnover, trade secrecy, and locating in countries with few serious competitors (Javorcik 2004a). Policies may be in place that mandate technology transfer to local firms and increase the transmission of knowledge and technology between the affiliate and the local market. Foreign investors may react, however, by limiting the level and nature of knowledge they transfer to affiliates in the first place.

Access to Finance
A few studies stress the role of financial markets in developing countries as a mediating factor for the absorption of FDI spillovers (Alfaro et al. 2010; Havranek and Irsova 2011). Multinationals can have a twofold effect on access to finance for local firms: (a) they may ease access to finance by bringing in scarce capital to developing countries; (b) if multinationals borrow from local financial institutions, they may increase local firms' financing constraints (Harrison, Love, and McMillan, 2004). This, in turn, can influence a local firm's absorptive capacity. Harrison, Love, and McMillan (2004) find evidence for the second effect using firm-level data for a cross-section of 38 high- and low-income countries. In this context, some countries have in the past attempted to limit access to local financial institutions for FDI firms in order to not crowd out scarce funds for domestic investments.

Other studies look explicitly at the role of financial development—representing access to finance—for FDI spillovers. Studies find that well-developed financial markets may facilitate a domestic firm's absorptive capacity. Agarwal, Milner, and Riaño (2011), for instance, find that FDI spillovers are lower or even negative for Chinese manufacturing firms that are credit-constrained. Javorcik and Spatareanu (2009) find that less liquidity-constrained firms are more likely to self-select into supplying multinationals. Havranek and Isrova (2011), in contrast, find evidence for lower FDI spillovers in more-developed financial systems. This implies that while access to finance in general should be easier for local suppliers, competition with foreign investors for limited financial resources might also increase with financial development, reducing local firms' absorptive capacity (second effect discussed).

Learning and Innovation Infrastructure
In the previous section "Domestic Firm Characteristics," we discussed why knowledge spillovers to domestic firms are more likely when workers in domestic firms possess the required skills and have access to specific information or knowledge (Paus and Gallagher 2008). The share of human capital at the firm level is influenced strongly by the local learning and innovation infrastructure, which includes the basic, higher, and vocational education system; the interaction between education and training institutions and the private sector; and the existence and embeddedness of technological research institutes. This infrastructure shapes the absorptive capacity of workers as well as, critically, managers. It also acts as an important determinant of how effectively the knowledge transmission

channels operate. These channels or "systems of innovation" (Lundvall 1992) operate at a national level and, importantly, have a strong local or regional dimension (Asheim and Isakesen 2002). This is important, as spillovers from innovation have been shown to decay rapidly with distance (Crescenzi, Rodriguez-Póse, and Storper 2007).

Meyer and Sinani (2009) discuss three measures of a country's human capital that significantly affect FDI spillovers: the share of workers with tertiary education, the R&D intensity in the private sector, and the number of patents per billion U.S. dollars granted to host country residents. They find a U-shaped effect—that is, spillovers increase only below or above certain threshold levels of human capital. Tytell and Yudaeva (2007) find evidence for Romania that productivity spillovers from FDI in manufacturing are significantly lower in regions with a low share of education. Monge-González and González-Alvarado (2007) illustrate the contribution of high-tech multinationals on skills development in Costa Rica, based on the experiences of Intel, Microsoft, and Cisco. They highlight the importance of the learning and innovation infrastructure; specifically, they describe how businesses, universities, and training and research institutes work together to develop curricula that better respond to the demands of the productive sector.

Trade, Investment, and Industrial Policy
A country's trade policy shapes the amount and type of foreign investment and, thus, affects the potential for FDI spillovers. Studies have related a country's trade policy regime to its capacity to attract foreign firms in the first place. Open trade regimes may be more likely to attract foreign investors than inward-oriented regimes, since they are less constrained by the size and efficiency of the local market in the first case (Crespo and Fontoura 2007). Foreign investors might also be more export-oriented in an open setting, increasing chances for local suppliers to become exporters, too. Moreover, foreign investors in an open trade setting are globally more integrated and therefore tend to adopt the newest technologies (Meyer and Sinani 2009). Others argue, however, that foreign investors are expected to bring new (or more developed) technologies to inward-oriented host countries in order to be successful, thus increasing the extent of positive spillovers through demonstration effects. Foreign investors in an outward-oriented trade policy regime, by contrast, tend to focus more strongly on international distribution and marketing and less on new technologies (Crespo and Fontoura 2007). In more protective countries, foreign firms may however be required to work more with local input suppliers as they cannot easily revert to importing.

Trade policy also affects domestic firms. Local firms in an open trade regime are more exposed to competitive pressures through international trade competition, which will prepare them to better absorb FDI spillovers (Meyer and Sinani 2009; Havranek and Irsova 2011). Moreover, a country's trade policy regime affects the likelihood of domestic firms becoming exporters and learning-by-exporting. Although the effect of exporting on domestic firms' absorptive capacities is ambiguous, exporting clearly moderates the direction and extent of FDI spillovers (see the previous section on domestic firm characteristics). Studies confirm that FDI spillovers are larger in countries that are more open to trade (Lesher and Miroudot 2008; Meyer and Sinani 2009; Havranek and Irsova 2011). Another recent study finds for China that both horizontal and vertical spillovers from FDI are negative when both final goods and input tariffs are higher (Du, Harrison, and Jefferson 2011). Studying horizontal FDI spillovers in the Indonesian manufacturing sector, Temenggung (2007) takes a different approach and focuses on different periods of economic openness rather than trade policy per se. The author finds that FDI spillovers were negative in the "pre-liberalization" period (1975–86), but positive during the "liberalization" period from 1986 up to the Asian crisis in 1996. These results confirm that openness in trade and FDI has a beneficial effect on the extent of FDI spillovers.

Investment policy and promotion also plays an important role in mediating spillovers, both in terms of contributing to attracting FDI in general (the focus of most export promotion efforts) and specific policies designed to promote spillovers between FDI and local firms (much less common). In a recent study using sectoral data across 105 developed and developing countries, Harding and Javorcik (2012) suggest that sectors that have been prioritized by investment promotion agencies in developing countries export relatively higher-quality products.[6] This suggests that investment promotion helps bring in firms that should have higher spillover potential (given their quality and technology position). Another study, looking at China's manufacturing sector (Du, Harrison, and Jefferson 2011), goes further to show a link with actual spillovers. The study finds that foreign firms enjoying investment subsidies generate positive backward spillovers, whereas foreign firms not enjoying such subsidies actually generate negative spillovers. In addition, this study finds that foreign firms enjoying tax exemption from value-added taxes generate higher forward spillovers than foreign firms that are not exempt from these taxes.

Special economic zones (SEZs) are a specific type of investment promotion tool that could affect the extent of FDI spillovers. Abraham, Konings, and Slootmaekers (2010) find for Chinese manufacturing firms that local firms located in SEZs have smaller productivity spillovers from FDI compared to domestic firms located in other places.[7] This might be because most SEZs focus on export processing combined with a high percentage of imported inputs, which limits the FDI potential, since the demand for local suppliers is constrained. Moreover, the spatial and legal structures that govern SEZs often inhibit their integration with the local economies that surround them.

Finally, industrial policies, particularly programs designed to support the development of local SMEs, can play an important role in mediating FDI spillovers. This may be particularly important where the technology and productivity gap is large between foreign and local firms, or if few local firms exist at all, due to a range of domestic market weaknesses. In the context of FDI spillovers, two policies have played an important role in facilitating spillovers in several sectors such as automotive and electronics: (a) collaboration with FDI firms, and (b) support for local supplier networks through supplier development programs, which are run by foreign affiliates but supported by governments. Local content requirements that demand a certain share of inputs to be sourced locally have also gained prominence, given the growth success of China and other emerging economies using such types of industrial policies. However, the track record of such approaches is mixed and depends critically on domestic absorption capacity and supplier development (see, for example, Greenaway 1992; Blomström and Wolff 1994).

Governance

Weak institutions—including corruption, red tape, or other inefficiencies or burdensome regulations—may prevent foreign and local investors from fully exploiting their competitive advantages. This may influence the type of FDI attracted in the first place, as well as the domestic firms' absorptive capacities. Empirical evidence is mixed. Using firm-level data for 17 emerging countries over the period 2002–05, Gorodnichenko, Svejnar, and Terrell (2007) find no evidence that the extent of FDI spillovers is affected by the degree of corruption (measured as bribes) or red tape (measured as managers' time spent with officials). Measuring transparency with a corruption perception index, Meyer and Sinani (2009) find evidence that a country's level of transparency has a U-shaped effect on FDI spillovers; that is, countries with a medium level of transparency benefit the least from FDI, while countries with a low and high level of transparency show stronger FDI spillovers. The impact also depends on the respective sectors. Moran (2006), for example, states that bribery and corruption are more commonly connected with foreign investors in infrastructure and extractive industries.

Development and implementation of effective FDI spillover policies in the areas discussed also depend critically on the quality of the host country's institutional environment. The potential for effective, proactive government policies in developing countries may be limited by capacity gaps. More importantly, policy may be impeded by lack of institutional contexts for close and effective relationships between the public and the private sector, as well as poor integration among FDI, SOEs, and domestic (formal and informal) enterprises (Pedersen and McCormick 1999). Trade and investment policies, in particular bilateral agreements, may also limit the policy space for more interventionist industrial policies, as has been discussed, for example, in the context of the European Union (EU) bilateral investment treaties.

Conclusions

Over the past 20 years, considerable attention has been paid to the role of FDI in development outcomes. A vast set of empirical evidence has been amassed, particularly over the past decade, on the existence and dynamics of FDI spillovers, the most important mechanism in the FDI-development nexus. While results are mixed, what is clear is that the benefits of FDI spillovers do not accrue automatically with the presence of FDI. Rather, a wide range of factors appear to mediate these outcomes.

The focus of research has shifted in recent years from identifying the existence of spillovers to understanding the conditions—the mediating factors—under which they are most likely to materialize. While this research is emerging quickly, important policy-relevant gaps remain. These include limited attention to the role of host country characteristics and, perhaps most notably, to how FDI dynamics and strategies—especially in the context of GVCs—mediate outcomes. Moreover, few studies have looked behind a narrow set of conditions to understand how mediating factors interact, and how they affect different transmission channels of spillovers. Besides transmission channels in value chains, research also needs to explore better the effect of changing market forces (demonstration and competition effects) and labor turnover. Acquiring this knowledge will help guide policies designed to remove barriers within transmission channels, enabling FDI spillover potential to translate into actual spillovers.

This chapter outlined a comprehensive conceptual framework for understanding the determinants of FDI spillovers, with an emphasis on mediating factors that shape the nature and extent of spillovers. Specifically, the chapter provided a detailed examination of three types of mediating factors: FDI spillover potential of foreign investors, the absorptive capacity of local agents (firms and workers), and the characteristics of the host country in which both are situated. The factors are assessed within a context of three different transmission channels: supply chains, labor turnover, and market restructuring.

Notes

1. The authors would like to thank Beata Javorcik for valuable comments.
2. See, for example, the research carried out under the "Making the Most of Commodities" program: http://commodities.open.ac.uk/mmcp.
3. An alternative explanation for the lower spillover effects could be the practice of "round tripping," where FDI from Taiwan, China; Hong Kong SAR, China; and Macao SAR, China, would, in reality, reflect disguised money from Chinese investors aiming to take advantage of FDI incentives in China (see, for example, World Bank 2002; Xiao 2004).
4. Pace is defined as the change of the number of foreign firms in a sector, province, and year compared to the previous year. Irregularity reflects the rhythm of building up foreign presence in a sector and province and is defined as the standard deviation of pace in every industry-province-year combination over the sample period (Wang et al. 2012).
5. Indeed, some econometric studies that attempt to estimate FDI spillovers on local firms' productivity prefer to use domestic firms' R&D share to capture technological backwardness rather than using productivity measures, as the latter may be correlated with the dependent variable.
6. As measured by unit values of exports.
7. However, looking at the effects by type of SEZ reveals that one particular SEZ among a group of six showed larger productivity spillovers from FDI compared to firms located elsewhere, namely the National Hi-Tech Industrial Development Zone. Unlike the other SEZs, this zone focuses on attracting foreign capital and technology.

References

Abraham, F., J. Konings, and V. Slootmaekers. 2010. "FDI Spillovers in the Chinese Manufacturing Sector: Evidence of Firm Heterogeneity." *Economies of Transition* 18: 143–82.

Agarwal, N., C. Milner, and A. Riaño. 2011. "Credit Constraints and FDI Spillovers in China." China and the World Economy Research Paper Series 21/2011, University of Nottingham.

Aitken, B., and A. Harrison. 1999. "Do Domestic Firms Benefit from Direct Foreign Investment? Evidence from Venezuela." *American Economic Review* 89 (3): 605–18.

Aitken, B., A. Harrison, and R. Lipsey. 1996. "Wages and Foreign Ownership: A Comparative Study of Mexico, Venezuela, and the United States." *Journal of International Economics* 40 (3–4): 345–71.

Alfaro, L., A. Chanda, S. Kalemli-Ozcan, and S. Sayek. 2010. "Does Foreign Direct Investment Promote Growth? Exploring the Role of Financial Markets on Linkages." *Journal of Development Economics* 91 (2): 242–56.

Almeida, R., and R. Aterido. 2011. "On-the-Job Training and Rigidity of Employment Protection in the Developing World: Evidence from Differential Enforcement." *Labour Economics* 18 (Supplement 1): S71–82.

Asheim, B., and A. Isaksen. 2002. "Regional Innovation Systems: The Integration of Local 'Sticky' and Global 'Ubiquitous' Knowledge." *The Journal of Technology Transfer* 27 (1): 77–86.

Balsvik, R. 2011. "Is Labor Mobility a Channel for Spillovers from Multinationals? Evidence from Norwegian Manufacturing." *Review of Economics and Statistics* 93 (1): 285–97.

Barrientos, S., and M. Visser. 2012. "South African Horticulture: Opportunities and Challenges for Economic and Social Upgrading in Value Chains." Working Paper 12, Capturing the Gains programme, School of Environment and Development, University of Manchester, U.K.

Barrios, S., L. Bertellini, and E. Strobl. 2006. "Coagglomeration and Spillovers." *Regional Science and Urban Economics* 36 (4): 467–81.

Barrios, S., S. Dimelis, H. Louri, and E. Strobl. 2004. "Efficiency Spillovers from Foreign Direct Investment in the EU Periphery: A Comparative Study of Greece, Ireland and Spain." *Weltwirtschaftliches Archiv* 140 (4): 688–705.

Barrios, S., and E. Strobl. 2002. "Foreign Direct Investment and Productivity Spillovers: Evidence from the Spanish Experience." *Review of World Economics* 138 (3): 459–81.

Blalock, G., and P. Gertler. 2009. "How Firm Capabilities Affect Who Benefits from Foreign Technology." *Journal of Development Economics* 90 (2): 192–99.

Blomström, M., and F. Sjöholm. 1999. "Technology Transfer and Spillovers: Does Local Participation with Multinationals Matter?" *European Economic Review* 43 (4–6): 915–23.

Blomström, M., and E. Wolff. 1994. "Multinational Corporations and Productive Convergence in Mexico." In *Convergence of Productivity: Cross National Studies and Historical Evidence*, edited by W. Baumol, R. Nelson, and E. Wolff, 263–83. Oxford, U.K.: Oxford University Press.

Braconier, H., K. Ekholm, and K. Knarvik. 2001. "In Search of FDI-Transmitted R&D Spillovers: A Study Based on Swedish Data." *Weltwirtschaftliches Archiv* 137 (4): 644–65.

Branstetter, L. 2006. "Is Foreign Direct Investment a Channel of Knowledge Spillovers? Evidence for Japan's FDI in the United States." *Journal of International Economics* 68 (2): 325–44.

Branstetter, L., R. Fisman, and C. Foley. 2006. "Do Stronger Intellectual Property Rights Increase International Technology Transfer? Empirical Evidence from U.S. Firm-Level Panel Data." *Quarterly Journal of Economics* 121 (1): 321–49.

Buckley, P., C. Wang, and J. Clegg. 2007. "The Impact of Foreign Ownership, Local Ownership and Industry Characteristics on Spillover Benefits from Foreign Direct Investment in China." *International Business Review* 16 (2): 142–58.

Castellani, D., and A. Zanfei. 2003. "Technology Gaps, Absorptive Capacity and the Impact of Inward Investments on Productivity of European Firms." *Economics of Innovation and New Technology* 12 (6): 555–76.

Cattaneo, O., G. Gereffi, and C. Staritz. 2011. *Global Value Chains in a Post-Crisis World.* Washington, DC: World Bank.

Crescenzi, R., A. Rodriguez-Póse, and M. Storper. 2007. "The Territorial Dynamics of Innovation: A Europe–United States Comparative Analysis." *Journal of Economic Geography* 7 (6): 673–709.

Crespo, N., and M. Fontoura. 2007. "Determinant Factors of FDI Spillovers—What Do We Really Know?" *World Development* 35 (3): 410–25.

Damijan, J., M. Knell, B. Majcen, and M. Rojec. 2003. "Technology Transfer through FDI in Top-10 Transition Countries: How Important Are Direct Effects, Horizontal and Vertical Spillovers?" William Davidson Working Paper 549, University of Michigan, Ann Arbor.

Dimelis, S., and H. Louri. 2002. "Foreign Ownership and Production Efficiency: A Quantile Regression Analysis." *Oxford Economic Papers* 54 (3): 449–69.

Du, L., A. Harrison, and G. Jefferson. 2011. "Do Institutions Matter for FDI Spillovers? The Implications of China's 'Special Characteristics'." NBER Working Paper 16767, National Bureau of Economic Research (NBER), Cambridge, MA.

Findlay, R. 1978. "Relative Backwardness, Direct Foreign Investment, and the Transfer of Technology: A Simple Dynamic Model." *Quarterly Journal of Economics* 92 (1): 1–16.

Fosfuri, A., M. Motta, and T. Ronde. 2001. "Foreign Direct Investment and Spillovers through Workers' Mobility." *Journal of International Economics* 53 (1): 205–22.

Fu, X. 2012. "Foreign Direct Investment and Managerial Knowledge Spillovers through the Diffusion of Management Practices." SLPTMD Working Paper Series 035, Department of International Development, University of Oxford.

Gentile-Lüdecke, S., and A. Giroud. 2012. "Knowledge Transfer from MNCs and Upgrading of Domestic Firms: The Polish Automotive Sector." *World Development* 40 (4): 796–807.

Gibbon, P. 2003. "Value-Chain Governance, Public Regulation and Entry Barriers in the Global Fresh Fruit and Vegetable Chain into the EU." *Development Policy Review* 21 (5): 615–25.

Girma, S. 2005. "Absorptive Capacity and Productivity Spillovers from FDI: A Threshold Regression Analysis." *Oxford Bulletin of Economics and Statistics* 67 (3): 281–306.

Girma, S., and H. Görg. 2007. "The Role of Efficiency Gap for Spillovers from FDI: Evidence from the U.K. Electronics and Engineering Sectors." *Open Economies Review* 18 (2): 215–32.

Girma, S., and K. Wakelin. 2007. "Local Productivity Spillovers from Foreign Direct Investment in the U.K. Electronics Industry." *Regional Science and Urban Economics* 37 (3): 399–412.

Giroud, A., B. Jindra, and P. Marek. 2012. "Heterogeneous FDI in Transition Economies: A Novel Approach to Assess the Developmental Impact of Backward Linkages." *World Development* 40 (11): 2206–20.

Glass, A., and K. Saggi. 2002. "Multinational Firms and Technology Transfer." *Scandinavian Journal of Economics* 104 (4): 495–513.

Godart, O., and H. Görg. 2013. "Suppliers of Multinationals and the Forced Linkage Effect: Evidence from Firm Level Data." CEPR Discussion Paper 9324, Centre for Economic Policy Research (CEPR), London.

Görg, H., and E. Strobl. 2005. "Spillovers from Foreign Firms through Worker Mobility: An Empirical Investigation." *Scandinavian Journal of Economics* 107 (4): 693–709.

Gorodnichenko, Y., J. Svejnar, and K. Terrell. 2007. "When Does FDI Have Positive Spillovers? Evidence from 17 Emerging Market Economies." IZA Discussion Paper 3079, Institute for the Study of Labor, Bonn, Germany.

Greenaway, D. 1992. "Trade Related Investment Measures and Development Strategy." *Kyklos* 45 (2): 139–60.

Greenaway, D., N. Sousa, and K. Wakelin. 2004. "Do Domestic Firms Learn to Export from Multinationals?" *European Journal of Political Economy* 20 (4): 1027–43.

Griffith, R., S. Redding, and H. Simpson. 2002. "Productivity Convergence and Foreign Ownership at the Establishment Level." Working Paper 22, Institute for Fiscal Studies, London.

Grünfeld, L. 2006. "Multinational Production, Absorptive Capacity, and Endogenous R&D Spillovers." *Review of International Economics* 14 (5): 922–40.

Hale, G., and C. Long. 2011. "Did Foreign Direct Investment Put an Upward Pressure on Wages in China?" *IMF Economic Review* 59: 404–30.

Harding, T., and B. Javorcik. 2012. "Foreign Direct Investment and Export Upgrading." *The Review of Economics and Statistics* 94 (4): 964–80.

Harrison, A. 1994. "Productivity, Imperfect Competition and Trade Reform." *Journal of International Economics* 36: 53–73.

Harrison, A., I. Love, and M. McMillan. 2004. "Global Capital Flows and Financing Constraints." *Journal of Development Economics* 75 (1): 269–301.

Havranek, T., and Z. Irsova. 2011. "Estimating Vertical Spillovers from FDI: Why Results Vary and What the True Effect Is." *Journal of International Economics* 85 (2): 234–44.

Hoekman, B., and B. Javorcik. 2006. "Lessons from Empirical Research on Technology Diffusion through Trade and Foreign Direct Investment." In *Global Integration and Technology Transfer*, edited by B. Hoekman and B. Javorcik. Washington, DC: Palgrave/World Bank.

Javorcik, B. 2004a. "Does Foreign Direct Investment Increase the Productivity of Domestic Firms? In Search of Spillovers through Backward Linkages." *American Economic Review* 94 (3): 605–27.

———. 2004b. "The Composition of Foreign Direct Investment and Protection of Intellectual Property Rights: Evidence from Transition Economies." *European Economic Review* 48 (1): 39–62.

———. 2008. "Can Survey Evidence Shed Light on Spillovers from Foreign Direct Investment?" *World Bank Research Observer* 23 (2): 139–59.

Javorcik, B., and M. Spatareanu. 2005. "Do Foreign Investors Care about Labor Market Regulations?" *Review of World Economics* 141 (3): 375–403.

———. 2008. "To Share or Not to Share: Does Local Participation Matter for Spillovers from Foreign Direct Investment?" *Journal of Development Economics* 85 (1–2): 194–221.

———. 2009. "Liquidity Constraints and Firms' Linkages with Multinationals." *World Bank Economic Review* 23 (2): 323–46.

———. 2011. "Does It Matter Where You Come From? Vertical Spillovers from Foreign Direct Investment and the Origin of Investors." *Journal of Development Economics* 96 (1): 126–38.

Jordaan, J. 2011. "Local Sourcing and Technology Spillovers to Mexican Suppliers: How Important Are FDI and Supplier Characteristics?" *Growth and Change* 42 (3): 287–319.

Kanturia, V. 2000. "Productivity Spillovers from Technology Transfer to Indian Manufacturing Firms." *Journal of International Development* 12 (3): 343–69.

———. 2001. "Foreign Firms, Technology Transfer and Knowledge Spillovers to Indian Manufacturing Firms: A Stochastic Frontier Analysis." *Applied Economics* 33 (5): 625–42.

———. 2002. "Liberalisation, FDI, and Productivity Spillovers—An Analysis of Indian Manufacturing Firms." *Oxford Economic Papers* 54 (4): 688–718.

Karpaty, P., and L. Lundberg. 2004. "Foreign Direct Investment and Productivity Spillovers in Swedish Manufacturing." Working Paper Series 194, Trade Union Institute for Economic Research (FIEF).

Katz, L. 1986. "Efficiency Wage Theories: A Partial Evaluation." In *NBER Macroeconomics Annual*, Vol. 1, edited by S. Fischer, 235–90. Cambridge, MA: MIT Press.

Keller, W., and S. Yeaple. 2009. "Multinational Enterprises, International Trade, and Productivity Growth: Firm Level Evidence from the United States." *The Review of Economics and Statistics* 91 (4): 821–31.

Kinoshita, Y. 2001. "R&D and Technology Spillovers through FDI: Innovation and Absorptive Capacity." CEPR Discussion Paper 2775, Centre for Economic Policy Research (CEPR), London.

Kokko, A. 1994. "Technology, Market Characteristics, and Spillovers." *Journal of Development Economics* 43 (2): 279–93.

Kokko, A., R. Tansini, and M. Zejan. 1996. "Local Technological Capability and Productivity Spillovers from FDI in the Uruguayan Manufacturing Sector." *Journal of Development Studies* 32 (4): 602–11.

Kugler, M. 2006. "Spillovers from Foreign Direct Investment: Within or Between Industries?" *Journal of Development Economics* 80 (2): 444–77.

Lall, S. 1980. "Vertical Inter-firm Linkages in LDCs: An Empirical Study." *Oxford Bulletin of Economics and Statistics* 42 (3): 203–26.

Lesher, M., and S. Miroudot. 2008. "FDI Spillovers and Their Inter-relationships with Trade." OECD Trade Policy Working Paper 80, Organisation for Economic Co-operation and Development (OECD), Paris.

Li, X., X. Liu, and D. Parker. 2001. "Foreign Direct Investment and Productivity Spillovers in the Chinese Manufacturing Sector." *Economic Systems* 25 (4): 305–21.

Lin, P., Z. Liub, and Y. Zhanga. 2009. "Do Chinese Domestic Firms Benefit from FDI Inflow? Evidence of Horizontal and Vertical Spillovers." *China Economic Review* 20 (4): 677–91.

Lundvall, B-Å., ed. 1992. *National Systems of Innovation: Towards a Theory of Innovation and Interactive Learning.* London: Pinter.

Meyer, K., and E. Sinani. 2009. "When and Where Does Foreign Direct Investment Generate Positive Spillovers? A Meta-Analysis." *Journal of International Business Studies* 40 (7): 1075–94.

Monge-González, R., and C. González-Alvarado. 2007. "The Role and Impact of MNCs in Costa Rica on Skills Development and Training: The Case of Intel Microsoft, and Cisco." Paper prepared for the International Labor Organization, Geneva, Switzerland.

Moran, T. 2006. "Harnessing Foreign Direct Investment for Development, Policies for Developed and Developing Countries." Center for Global Development, Washington, DC.

Morris, M., R. Kaplinsky, and D. Kaplan. 2011. "Commodities and Linkages: Meeting the Policy Challenge." MMCP Discussion Paper 14, Making the Most of Commodities Programme (MMCP), The Open University, U.K.

Morris, M., C. Staritz, and J. Barnes. 2011. "Value Chain Dynamics, Local Embeddedness, and Upgrading in the Clothing Sectors of Lesotho and Swaziland." *International Journal of Technological Learning, Innovation and Development* 4 (1–3): 96–119.

Nadvi, K., and H. Schmitz. 1994. "Industrial Clusters in Less Developed Countries: A Review of Experiences and Research Agenda." Discussion Paper 339, Institute of Development Studies, University of Sussex, Brighton, U.K.

Pack, H., and K. Saggi. 2001. "Vertical Technology Transfer via International Outsourcing." *Journal of Development Economics* 65 (2): 389–415.

Paus, E., and K. Gallagher. 2008. "Missing Links: Foreign Investment and Industrial Development in Costa Rica and Mexico." *Studies of Comparative International Development* 43 (1): 53–80.

Pedersen, P., and D. McCormick. 1999. "African Business Systems in a Globalizing World." *The Journal of Modern African Studies* 37 (1): 109–35.

Phillips, R., and J. Henderson. 2009. "Global Production Networks and Industrial Upgrading: Negative Lessons from Malaysian Electronics." *Austrian Journal for Development Studies* 25 (2): 38–61.

Pigato, A. 2001. "The Foreign Direct Investment Environment in Africa." Africa Region Working Paper Series 15, The World Bank, Washington, DC.

Plank, L., and C. Staritz. 2013. "'Precarious Upgrading' in Electronics Global Production Networks in Central and Eastern Europe: The Cases of Hungary and Romania." Working Paper 41, ÖFSE, Vienna.

Ponomareva, N. 2000. "Are There Positive or Negative Spillovers from Foreign-Owned to Domestic Firms?" Working Paper BSP/00/042, New Economic School, Moscow.

Quddus, M. and S. Rashid. 2000. *Entrepreneurs and Economic Development: The Remarkable Story of Garment Exports from Bangladesh.* Dhaka, Bangladesh: University Press Limited.

Resmini, L., and M. Nicolini. 2007. "Productivity Spillovers and Multinational Enterprises: In Search of a Spatial Dimension." Working Paper 10, Dynamic Regions in a Knowledge-Driven Global Economy: Lessons and Policy Implications for the EU (DYNREG).

Rodriguez-Clare, A. 1996. "Multinationals, Linkages, and Economic Development." *The American Economic Review* 86 (4): 852–73.

Rodrik, D. 2003. "Appel Inaugural Lecture." Columbia University, March 27.

Saggi, K. 2002. "Trade, Foreign Direct Investment, and International Technology Transfer: A Survey." *The World Bank Research Observer* 17 (2): 191–235.

Schoors, K., and B. van der Tol. 2002. "Foreign Direct Investment Spillovers within and between Sectors: Evidence from Hungarian Data." Working Paper 2002/157, University of Gent.

Sinani, E., and K. Meyer. 2004. "Spillovers of Technology Transfer from FDI: The Case of Estonia." *Journal of Comparative Economics* 32 (3): 445–66.

Sjöholm, F. 1999. "Productivity Growth in Indonesia: The Role of Regional Characteristics and Direct Foreign Investment." *Economic Development and Cultural Change* 47 (3): 559–84.

Smeets, R. 2008. "Collecting the Pieces of the FDI Knowledge Spillovers Puzzle." *World Bank Research Observer* 23 (2): 107–38.

Staritz, C. 2011. *Making the Cut? Low-Income Countries and the Global Clothing Value Chain in a Post-Quota and Post-Crisis World.* Washington, DC: World Bank.

Staritz, C., and M. Morris. 2012. "Local Embeddedness, Upgrading and Skill Development: Global Value Chains and Foreign Direct Investment in Lesotho's Apparel Industry." Working Paper 32, ÖFSE, Vienna.

———. 2013. "Local Embeddedness and Economic and Social Upgrading in Madagascar's Export Apparel Industry." Working Paper 38, ÖFSE, Vienna.

Suyanto, S., and R. Salim. 2010. "Sources of Productivity Gains from FDI in Indonesia: Is It Efficiency Improvement or Technological Progress?" *The Developing Economies* 48 (4): 450–72.

Takii, S. 2005. "Productivity Spillovers and Characteristics of Foreign Multinational Plants in Indonesian Manufacturing, 1990–1995." *Journal of Development Economics* 76 (2): 521–42.

Temenggung, D. 2007. "Productivity Spillovers from Foreign Direct Investment: Indonesian Manufacturing Industry's Experience 1975–2000." Dynamics, Economic Growth, and International Trade (DEGIT) Conference Papers, Kiel Institute for the World Economy.

Thompson, E. 2002. "Clustering of Foreign Direct Investment and Enhanced Technology Transfer: Evidence from Hong Kong Garment Firms in China." *World Development* 30 (5): 873–89.

Tondl, G., and J. Fornero. 2010. "Sectoral Productivity and Spillover Effects of FDI in Latin America." FIW Working Paper 53, Austrian Institute of Economic Research (WIFO).

Torlak, E. 2004. "Foreign Direct Investment, Technology Transfer, and Productivity Growth in Transition Countries—Empirical Evidence from Panel Data." Cege Discussion Paper 26, Center of Globalization and Europeanization of the Economy (Cege), University of Göttingen, Department of Economics.

Toth, I., and A. Semjen. 1999. "Market Links and Growth Capacity of Enterprises in a Transforming Economy: The Case of Hungary." In *Market Links, Tax Environment and*

Financial Discipline of Hungarian Enterprises, edited by I. Toth and A. Semjen, 1–37. Budapest: Institute of Economics, Hungarian Academy of Sciences.

Tytell, I., and K. Yudaeva. 2007. "The Role of FDI in Eastern Europe and New Independent States: New Channels for the Spillover Effect." In *Foreign Direct Investment in Europe: A Changing Landscape*, edited by K. Liebscher, J. Christl, P. Mooslechner, and D. Ritzberger-Grünwald, 76–86. Cheltenham, U.K.: Edward Elgar.

Wang, J., and M. Blomström. 1992. "Foreign Investment and Technology Transfer: A Simple Model." *European Economic Review* 36 (1): 137–55.

Wang, C., Z. Deng, M. Kafouros, and Y. Chen. 2012. "Reconceptualizing the Spillover Effects of Foreign Direct Investment: A Process-Dependent Approach." *International Business Review* 21 (3): 452–64.

Weiss, A. 1991. *Efficiency Wages*. Princeton, NJ: Princeton University Press.

World Bank. 2002. "Box 2.3: Round-Tripping of Capital Flows between China and Hong Kong." In *Global Development Finance: Financing the Poorest Countries*. Washington, DC: World Bank.

Xiao, G. 2004. "People's Republic of China's Round-Tripping FDI: Scale, Causes and Implications." Asian Development Bank Institute Discussion Paper 7, July. Asian Development Bank Institute, Tokyo.

Yellen, J. 1984. "Efficiency Wage Models of Unemployment." *American Economic Review* 74 (2): 200–25.

Yudaeva, K., K. Kozlov, N. Malentieva, and N. Ponomareva. 2003. "Does Foreign Ownership Matter? The Russian Experience." *Economics of Transition* 11 (3): 383–409.

Zhang, Y., H. Li, Y. Li, and L. Zhou. 2010. "FDI Spillovers in an Emerging Market: The Role of Foreign Firms' Country Origin Diversity and Domestic Firms' Absorptive Capacity." *Strategic Management Journal* 31 (9): 969–89.

PART 2

Quantitative Studies

The Role of Mediating Factors for FDI Spillovers in Developing Countries: Evidence from a Global Dataset   59

Determining the Nature and Extent of Spillovers: Empirical Assessment   87

CHAPTER 3

The Role of Mediating Factors for FDI Spillovers in Developing Countries: Evidence from a Global Dataset

Thomas Farole and Deborah Winkler[1,2]

Abstract

Using a cross-section of more than 25,000 domestic manufacturing firms in 78 low- and middle-income countries from the World Bank's Enterprise Surveys over the period 2006–10, we assess how mediating factors (MFs) influence productivity spillovers to domestic firms from foreign direct investment (FDI). We differentiate between three types of MFs: (a) a foreign investor's spillover potential, (b) a domestic firm's absorptive capacity, and (c) a country's institutional framework. We find that all three affect the extent and direction of FDI spillovers on domestic firm productivity. Moreover, we find that the impact of MFs depends on domestic firms' productivity and the structure of foreign ownership.

Introduction

A vast set of empirical evidence has been amassed over the past decade on the existence and direction of foreign direct investment (FDI)–generated horizontal and vertical spillovers (for a review of the literature, see, for example, Görg and Greenaway 2004; Lipsey and Sjöholm 2005; Smeets 2008; Havranek and Irsova 2011). Overall, the results are mixed, and suggest that the theoretical spillover effects often do not automatically materialize just because a country is able to attract FDI in the first place. As a result, more and more research has been devoted to understanding the various conditions that may explain these mixed results. Three major types of mediating factors (MFs) have been identified: (a) characteristics of foreign firms, which mediate spillover potential; (b) characteristics of domestic firms, which mediate absorptive capacity to internalize spillovers; and (c) differences in host country factors (Castellani and Zanfei 2003; Lipsey and Sjöholm 2005), which mediate both domestic and foreign firm characteristics as well as the transmission channels for spillovers (Paus and Gallagher 2008).

Using a cross-section of more than 25,000 domestic manufacturing firms in 78 low- and middle-income countries (LMICs) from the World Bank's Enterprise Surveys Indicator Database, we assess how MFs influence productivity spillovers to domestic firms from FDI. This chapter contributes to the growing body of research on MFs for FDI spillovers in several ways.

First, most studies are limited to examining a single mediating factor, with the majority focusing on the absorptive capacity of domestic firms. Fewer studies analyze the role of foreign investors or host country characteristics and the institutional context for FDI-enhanced productivity spillovers. To our knowledge, only Havranek and Irsova (2011) control for all three types of mediating factor. Their meta-analysis uses the t-statistic of existing FDI spillover estimates from other studies as the dependent variable. In contrast, this chapter estimates the impact of FDI directly on productivity, and introduces all three types of MFs in the form of interaction terms with the FDI spillover variable, which—to our knowledge—has not been done before. While the methodological approach of this chapter is similar to Blalock and Gertler (2009), their study focuses only on a domestic firm's absorptive capacities.

Second, related to the relatively low number of studies that take into account the characteristics of foreign investors is the fact that studies have neglected the specific dynamics within global value chains (GVCs). Global production networks are "led by large firms based typically in the industrialized countries, and relying often on complex networks of suppliers around the world" (Milberg and Winkler 2013, 10). The potential for FDI spillovers, however, is determined by the GVC in which foreign firms operate and by the specific GVC dynamics, including the FDI motive and sourcing strategy, among others (Paus and Gallagher 2008). This study addresses this gap in the literature by including two measures that proxy for foreign investors' FDI motive and sourcing behavior.

Third, most studies adding MFs focus on one or a few variables only. Even the meta-analysis by Meyer and Sinani (2009) covers only seven institutional variables.[3] This chapter takes a more comprehensive approach, focusing on four variables reflecting the FDI spillover potential, six variables representing a domestic firm's absorptive capacity, and eleven variables covering national characteristics and the institutional framework.

Fourth, as mentioned, most studies—with the exception of meta-analyses (for example, Meyer and Sinani 2009; Havranek and Irsova 2011)—tend to focus on a specific country setting. Such studies have the advantage of examining a specific locational context. However, using a cross-section of 78 LMICs allows us to study the MFs at a more general level and avoid the risk of country bias.

Fifth, it might be possible that MFs, such as firm size or exporting capabilities, are a reflection of heterogeneous firm-level productivity (Melitz 2003). Girma and Görg (2007, 220), for example, point out that "[i]n the presence of heterogeneous productivity processes, it is more appropriate (and arguably more interesting) to examine the dynamics of productivity at different points of the distribution rather than 'average' properties (that is conditional means)." Acknowledging the fact that firms are heterogeneous in terms of their productivity, we also examine if the role of MFs for FDI spillovers is a function of domestic firms' productivity.

Sixth, studies have pointed to the higher spillover potential of foreign affiliates with partial foreign ownership (Javorcik 2004; Javorcik and Spatareanu 2008; Abraham, Konings, and Slootmaekers 2010; Havranek and Irsova 2011). They confirm the view that the likelihood of technology leakages and knowledge spillovers are higher from foreign firms with local participation. To our knowledge, this chapter is the first to show that foreign ownership structure not only matters for FDI spillovers, but also for the impact of MFs on domestic firm productivity.

And finally, we apply an instrumental variables approach to address the potential endogeneity between FDI spillovers and domestic firm productivity. Foreign firms may be attracted into a specific sector in a country because of some unobserved characteristics that are correlated with domestic firm productivity. Most FDI spillover panel studies include fixed country-sector effects to control for such unobservable effects. However, only few studies take the alternative approach and use instruments for FDI spillovers (see Haskel, Pereira, and Slaughter 2007; Keller and Yeaple 2009; Jordaan 2011a). In this chapter, we use three different instruments for the FDI spillover variable.

This chapter is structured as follows. The next section introduces the data and econometric model. In the third section we present our regression results, while the fourth section concludes.

Empirical Model

Spillovers through Supply Chains

Following Blalock and Gertler (2009), we define the following equation:

$$\ln prod_{irst} = \alpha_0 + \beta FDI_{sct} + \gamma FDI_{sct} * MF + D_r + D_s + D_t + \varepsilon_{irst} \qquad (3.1)$$

where subscript i stands for firm, r for (subnational) region, s for sector, c for country, and t for year. The term α_0 designates the constant, D_r region fixed effects, D_s sector fixed effects, D_t year fixed effects, and ε_{irst} the idiosyncratic error term. The term $\ln prod$ is a measure of productivity in logarithms, FDI a measure of FDI spillovers at the sectoral level in a country, and $FDI_{sct} * MF$ the interaction term of FDI with a mediating factor MF. The total effect of FDI on productivity is given by $\beta + \gamma MF$. Our coefficient of interest is γ. The total effect of FDI on productivity will be larger (smaller, respectively) than β if the coefficient of the interaction term is positive (negative, respectively); that is, $\gamma > 0$ ($\gamma < 0$, respectively).

Following the literature, we use the share of foreign output as a percentage of total output at the sectoral level in a country as our measure of intra-industry FDI spillovers.[4]

$$FDI^Y_{sct} = \frac{\sum_{i \in sct} Y_i^{for}}{\sum_{i \in sct} Y_i} \qquad (3.2)$$

where $i \in sct$ indicates a firm in a given sector s of country c at time t, Y_i is firm-level output in a given sector of country c at time t, and Y_i^{for} is output if the firm is foreign. As is common in the literature on FDI, we only consider firms as foreign with a foreign ownership of 10 percent or higher.

This measure of intra-industry FDI spillovers in the strict sense captures only horizontal spillovers. However, since sectors are broadly defined (see annex 3B), FDI spillovers are likely to capture some vertical spillovers. For example, "auto and auto components" includes both final automotive producers and suppliers of automotive components—thus, FDI in this sector could affect both domestic final producers of cars and domestic suppliers of auto components. Similar situations are also likely in sectors such as food, electronics, and chemicals and pharmaceuticals.

To capture firm-level productivity, we use labor productivity (LP), defined as value added per worker. LP measures are only available for manufacturing sectors.[5] Since LP is mainly determined by capital intensity, we add firm-level capital intensity, defined as capital stock per worker,[6] in logarithms as an explanatory variable to each specification in order to avoid an omitted variable bias. This yields the following estimation equation:

$$\ln lp_{irst} = \alpha_0 + \beta FDI^Y_{sct} + \gamma FDI^Y_{sct} * MF + \ln capint_{irst} + D_r + D_s + D_t + \varepsilon_{irst} \qquad (3.3)$$

Since we are only interested in spillovers from foreign to domestic firms, we will run the regressions across domestic firms only. In a first step, we will run the baseline regressions as specified in equation (3.3) using ordinary least squares (OLS).

The cross-sectional nature of our dataset (see the later section "Measures of Foreign Spillover Potential"), however, renders the identification of a causal relationship between FDI spillovers and domestic firm productivity difficult. Foreign firms may be attracted into a specific sector in a country because of some unobserved characteristics that are correlated with domestic firm productivity. Aitken and Harrison (1999) find that controlling for such unobservable effects at the country-sector level leads to negative spillovers, while not controlling for fixed country-specific sector effects reverses the impact.

To control for such effects adequately, we would need to include fixed country-sector effects, which is not possible using cross-sectional data, as those would be perfectly correlated with the spillovers variable. One alternative approach is to instrument for FDI spillovers. Only few studies used instruments to address the potential endogeneity between FDI spillovers and domestic firm productivity. Haskel, Pereira, and Slaughter (2007) instrument their sectoral FDI spillover variable for the United Kingdom with sectoral inward FDI data from the United States. Similarly, in his study on Mexican manufacturing firms, Jordaan (2011a) uses the U.S. sectoral foreign employment share as instrument for his sectoral FDI spillover measure.

In their study on FDI spillovers in the United States, Keller and Yeaple (2009) use contemporaneous changes in shipping costs and tariffs and lagged levels of the real exchange rate interacted with industry dummies to instrument for sectoral FDI spillovers and imports. Studying the relationship between FDI and export upgrading, Harding and Javorcik (2011) instrument for inward FDI flows and stock to the United States by using information on industry-level targeting of investment promotion agencies. In a second step, we will estimate the regressions using an instrumental variables two-stage least squares (IV 2SLS) approach.

Data and Instruments

The World Bank Enterprise Analysis Unit published in 2011 the "Enterprise Surveys Indicator Database" (World Bank 2011).[7] This publication covers 215 Enterprise Surveys for 126 countries over the period 2006–10. Enterprise Surveys represent a comprehensive source of firm-level data in emerging markets and developing economies. One major advantage of Enterprise Surveys is that the survey questions are the same across all countries. Moreover, the Enterprise Surveys represent a random sample of firms using three levels of stratification: sector, firm size, and region. Sectors are based on the International Standard Industrial Classification (ISIC) Rev. 3.1 classification, but in some cases are further aggregated.

The Enterprise Surveys Indicator Database covers a wide range of indicators on firm characteristics, the business environment, innovation and technology, and workforce and skills, among others. We merged this dataset with data on firm-level output, value added, and capital stock obtained from the Enterprise Analysis Unit of the World Bank.[8] All local currencies have been converted into U.S. dollars and deflated using a gross domestic product (GDP) deflator in U.S. dollars (base year 2000). Exchange rates and GDP deflators have been obtained from the World Development Indicators (WDI).[9]

The following rules are applied to the dataset: (a) we include only the most recent Enterprise Surveys for each country; (b) we include only countries that cover foreign firms in the surveys;[10] (c) we drop high-income countries to cover only emerging or developing countries;[11] and (d) we drop countries for which we cannot calculate our FDI spillover measure due to unavailable output data. We focus only on the effects of productivity spillovers on domestic manufacturing firms, since productivity measures were unavailable for services firms.

This procedure results in more than 25,000 domestic firms and 3,400 foreign firms in 78 LMICs covering 11 manufacturing sectors. The list of countries, year of most recent survey, and number of domestic and foreign manufacturing firms by country can be found in annex 3A.

We combine these firm-level data with country-level data to control for national characteristics, including a country's institutional framework. The data source for each variable is indicated later in the section "Measures of National Characteristics and Institutions." Data for the national variables are aligned with the survey year of a country's Enterprise Survey (see annex 3A for information on survey years). In a few cases where national data were unavailable for the specific survey year, we choose the observation of the nearest available year.

We include three different instruments for FDI spillovers. The selection of these instruments is based on the assumption that the instruments are correlated with the FDI spillover variable, but not with domestic firm productivity in the country of interest. First, we take advantage of the fact that for many countries the Enterprise Surveys Indicator Database publishes surveys for more than one year. Based on the previous year for such countries, we are able to calculate the sectoral spillover variable as defined in equation (3.2) to be used as an instrument. For some countries we have to rely on the previous version of the Enterprise Surveys Indicator Database, which covers the period 2002–05.[12] In most cases, the time lag between the original and lagged spillover variables covers between two and four years. Therefore, it is safe to assume that the instrument is correlated with the spillover variable in t, but not correlated with domestic firm productivity in t.

Second, we add a sectoral measure of sector targeting by investment promotion agencies, which is similar to the approach of Harding and Javorcik (2011).[13] The measure equals 1 if a sector has been targeted by a country's investment promotion agency in a certain year and 0 if not. As our spillover measure reflects FDI presence rather than FDI flows, we aggregate the dummies over the period 2000–04 to obtain a measure of total length of sector targeting in the early 2000s. The sum can range from 0 (no targeting over this period) to 5 (continuous targeting over this period). To control for nonlinearities, we use the measure in natural logarithms, that is, ln(total sector targeting + 1). We believe that total sector targeting is well suited as an instrument, as it is correlated with the FDI spillover variable, but not with domestic firm productivity—especially since there is a time lag between these two measures.

Finally, following Keller and Yeaple (2009), we include sectoral tariffs obtained from the World Bank's World Integrated Trade Solution Database.[14] The data are based on the nonagricultural market access method of the World Trade Organization (WTO) (effectively applied rates).[15] Since some of our sectors are more aggregated than the original ISIC Rev. 3.1 classification, we aggregate tariffs up using import shares as weights. Tariff data are aligned with the survey year of a country's Enterprise Survey. In a few cases where tariff data are unavailable for the specific survey year, we choose the observation of the nearest available year before the survey year.

Measures of Foreign Spillover Potential

We now turn to the MFs. In a first step, we present measures of FDI spillover potential at the sectoral level based on available data in the Enterprise Surveys Indicator Database. All of the following variables are averages across all foreign firms within a specific sector of a country.

- *own* = a sector's average percentage of foreign ownership in a country.
- *market* = a sector's average percentage of FDI sales to the domestic market in a country. This measure serves as a proxy for a sector's average FDI motive in a country, whereby a higher share is associated with more market-oriented FDI.
- *inp* = a sector's average percentage of domestic input purchases of FDI firms in a country. This measure captures a sector's average sourcing strategy of foreign firms in a country, whereby a higher share is associated with more local sourcing.
- *tech* = *iso* + *tech_for* + *website* + *email* with $0 \leq tech \leq 4$, where *iso* = 1 if firm owns internationally recognized quality certification and 0 otherwise, *tech_for* = 1 if firm uses technology licensed from foreign firms and 0 otherwise, *website* = 1 if firm uses own website to communicate with clients or suppliers, *email* = 1 if firm uses email to communicate with clients or suppliers. The technology indicator serves as a proxy for a sector's average FDI technology intensity in a country.

Measures of Absorptive Capacity

In a second step, we include the following measures of absorptive capacity that were available in the Enterprise Surveys Indicator Database.

- *gap* = domestic firm's LP relative to median LP of multinational firms in sector in natural logarithms.
- *tech* = domestic firm's technology indicator as defined in previous section, where $tech \in \{0, 1, 2, 3, 4\}$. The technology indicator serves as a proxy for research and development (R&D) intensity, which is unavailable.
- *skills* = domestic firm's share of high-skilled labor in firm's total labor force.
- *size* = domestic firm's total number of permanent and temporary employees in natural logarithms.
- *aggl* = region's total number of manufacturing and services firms as percentage of a country's total number of manufacturing and services firms. This measure is a proxy for urbanization economies (locational advantages) and covers both domestic and foreign firms.[16]
- *exp* = domestic firm's share of direct or indirect exports in firm sales.

Measures of National Characteristics and Institutions

In a third step, we turn to MFs at the national level. We include the following country-level variables into the model.

- *labor* = measure of labor freedom in natural logarithms from the Heritage Foundation and captures labor market institutions. The variable ranges from 0 to 100 (highest labor freedom) and includes various aspects of the legal and regulatory framework of a country's labor market, such as minimum wages; laws inhibiting layoffs; severance requirements; and measurable regulatory burdens on hiring, hours, and so forth. The measure is mainly based on data from the World Bank's Doing Business database.
- *finance* = measure of financial freedom in natural logarithms from the Heritage Foundation. The variable measures banking efficiency as well as a measure of independence from government control and interference in the financial sector, with scores ranging from 0 to 100 (highest financial freedom). This measure relies on various underlying data sources, including (in order of priority) the Economist Intelligence Unit (EIU), the International Monetary Fund, the Organisation for Economic Co-operation and Development, and official government publications of each country, among others.
- *educ1* = government spending on education as percentage of GDP from the WDI database.[17]
- *educ2* = people with completed secondary and tertiary education as percentage of population aged 15 and over from Barro and Lee (2010).
- *rd* = a country's expenditures on R&D as percentage of GDP from the WDI database.
- *investment* = measure of investment freedom in natural logarithms from the Heritage Foundation and serves as a proxy for investment promotion. The score ranges from 0 to 100 (highest investment freedom) and measures the ability of individuals and firms to move their resources in and out of specific activities both internally and across the country's borders. This variable is mainly based on official government publications of each country on capital flows and foreign investment.
- *trade1* = a country's share of exports of goods and services as percentage of GDP from the WDI database.
- *trade2* = measure of trade freedom in logarithms from the Heritage Foundation, which is a composite measure of the trade-weighted average applied tariff rate and nontariff barriers with scores ranging from 0 to 100 (highest trade freedom), reflecting the absence of trade protectionism. The measure is based on various underlying sources including data from World Bank, the WTO, and the EIU, among others.
- *business* = measure of business freedom in natural logarithms from the Heritage Foundation as an outcome-based indicator of a country's institutional

development. It is a measure reflecting the ability to start, operate, and close a business with scores ranging from 0 to 100 (highest business freedom). The measure mainly relies on the World Bank's Doing Business study.
- *hhi* = Herfindahl-Hirschman Index (HHI) of sector concentration to capture competition in a domestic firm's sector. The HHI of sector concentration is defined as the sum of squares of a firm's output share by sector. If only one firm operates in a sector, the HHI would be 1. A lower HHI reflects a higher sectoral diversity. This measure includes both domestic and foreign firms.
- *income* = a country's per capita GDP (U.S. dollars at 2000 prices) in natural logarithms from the World Bank's WDI database. It captures national competition, as well as other aspects of the national and institutional environment.

Regression Results

FDI Spillover Potential
Overall Results

In the following, we apply OLS and IV 2SLS regressions. All regression results follow equation (3.1) and include sector, subnational region, and year fixed effects. Standard errors are robust to heteroscedasticity and clustered at the country-sector level. FDI spillovers are defined as in equation (3.2). The summary statistics are reported in annex 3C. Our data sample only includes domestic manufacturing firms, that is, all firms with a foreign ownership share of less than 10 percent (see annex 3A). Table 3.1 shows how the FDI spillover potential influences domestic firm productivity. We only report the results for the interaction terms of foreign ownership with foreign firm characteristics (for a full set of regression results, see Farole and Winkler 2012). Foreign firm characteristics are averages across all foreign firms within a specific sector of a country. Column A reports the OLS regression results, while column B shows the IV 2SLS regression results.[18] Each column-row combination refers to one separate regression. As could be expected, capital intensity shows a positive and significant effect on LP (not shown in table 3.1).

Horizontal FDI spillovers mostly show a negative effect on LP of domestic firms, although they are only significant in column A, row (2a) (not shown in table 3.1). What may explain the negative FDI spillovers? In the short term, local firms might face losses in their market share due to increased competition (competition effect). This is likely to affect high-productivity firms more (as they are more likely to be in direct competition with FDI in both domestic and export markets), and may explain why our results show a stronger impact for high-productivity firms. Loss of market share, in turn, requires them to produce at higher average unit costs, as a declining amount of output shifts them left on the economies-of-scale curve (Harrison 1994; Aitken and Harrison 1999; Crespo and Fontoura 2007). It is also more likely in the short term that foreign firms will bid away high-quality labor from domestic firms by offering higher wages

Table 3.1 FDI Spillover Potential, FDI Spillovers

		A	B	C	D	E	F	G
	Estimator:	OLS	IV 2SLS	IV 2SLS	IV 2SLS	IV 2SLS	IV 2SLS	IV 2SLS
Interaction term with FDI spillovers	Productivity level:	All	All	Low	Medium	High	All	All
	Foreign ownership:	All	All	All	All	All	Full	Partial
(1)	own_{sct}	0.3158	0.5961	1.4043	−0.8388	−0.1703		0.1544
		(0.228)	(0.680)	(0.469)	(0.601)	(0.938)		(0.938)
(2a)	$market_{sct}$	0.3979**	0.8586	1.9283**	1.2940*	−0.1809	3.6360	0.8028
		(0.028)	(0.141)	(0.018)	(0.078)	(0.868)	(0.363)	(0.321)
(2b)	$market_{sct}$		2.1830	5.2864**	5.0172**	0.5402	7.0625*	1.2382
			(0.123)	(0.035)	(0.025)	(0.760)	(0.098)	(0.494)
(3a)	inp_{sct}	−0.0844	1.1346	2.2108**	1.3829*	0.4191	−0.1213	2.3147**
		(0.700)	(0.114)	(0.038)	(0.070)	(0.677)	(0.889)	(0.039)
(3b)	inp_{sct}		4.0729	6.7001	4.7339	1.7521	−0.3045	1.5365
			(0.198)	(0.152)	(0.171)	(0.657)	(0.852)	(0.339)
(4)	$tech_{sct}$	−0.1050	−0.1177	−0.0793	−0.0700	−0.2219	−0.1679	−0.2242
		(0.147)	(0.479)	(0.750)	(0.728)	(0.385)	(0.212)	(0.721)

Source: Farole and Winkler 2012.
Note: Each column-row combination designates a separate regression. The dependent variable is labor productivity $lnlp_{first}$. All regressions control for the foreign ownership share in a sector and country and capital intensity at the firm level and include sector, subnational region, and year fixed effects (see equation 3.3). Standard errors are clustered at the country-sector level. Low productivity: $LP \leq 33.33$ percentile. Medium productivity: 33.33 percentile $< LP \leq 66.67$ percentile. High productivity: $LP > 66.67$ percentile. Full foreign ownership: foreign participation = 100 percent. Partial foreign ownership: 10 percent $\geq$ foreign participation $<$ 100 percent. Besides the three instruments described in the section "Data and Instruments," rows (2b) and (3b) include a fourth instrument for the interaction term of foreign direct investment (FDI) spillovers with *market* (2b) and *inp* (3b) (see description in text). *$p < 0.1$, **$p < 0.05$, ***$p < 0.01$ (p values in parentheses). For the full set of regression results, see Farole and Winkler (2012).

and benefits (labor turnover effect), resulting in a potentially negative spillover effect (Sinani and Meyer 2004; Hoekman and Javorcik 2006; Crespo and Fontoura 2007).

Both short-term effects can reverse in the medium to long term, if domestic firms become more productive due to increased competition and start to absorb skilled workers from multinationals. Research confirms that new entrants can have a Schumpeter winds-of-destruction impact, leading to a beneficial restructuring of the entire industry over time, including opportunities for better-performing local participants and local suppliers (see, for example, Iacovone et al. 2011). Given that our cross-sectional data sample only allows us to control for short-term effects, the negative coefficients on the FDI spillover variable are not surprising. Moreover, they are in line with other studies rejecting the existence of positive intra-industry FDI spillovers (Havranek and Irsova 2011) or studies that find negative intra-industry spillovers (for example, Aitken and Harrison 1999; Djankov and Hoekman 2000; Konings 2001, among others).[19]

A sector's average share of foreign ownership in a country (*own*) shows a positive coefficient sign, but is insignificant (columns A and B, row 1). The interaction terms with a sector's average percentage of FDI sales to the domestic market in a country (*market*) has a significantly positive effect on domestic firms' productivity in the OLS regressions (column A, row 2a), confirming our hypothesis that market-seeking FDI is more likely not only to provide a higher spillover potential, but also to translate into more positive actual spillovers.

Since the *market* variable might be influenced by the productivity of domestic competitors, we additionally instrument for the interaction term of FDI spillovers with the *market* variable in column B, row (2b). We choose the sectoral average number of days to clear direct exports through customs in a country (in natural logarithms) from the Enterprise Surveys Indicator Database as an inverse instrument. One could argue that longer waiting times at customs motivate foreign firms to rely more strongly on the host market for sales. The coefficient signs in the IV 2SLS regressions are also positive, but insignificant (column B, rows 2a and 2b). However, the Hansen J-test of overidentifying restrictions (not shown in the table) implies that the instruments are not accepted in these specifications.

Analogously, a sector's average percentage of domestic input purchases of FDI firms in a country (*inp*) might be endogenous, as the sectoral share of inputs purchased locally might be influenced by domestic firm productivity. We therefore additionally instrument for the interaction term of FDI spillovers with the *inp* variable in column B, row (3b), using the sectoral average number of days to clear imports through customs in a country (in natural logarithms) from the Enterprise Surveys Indicator Database as an inverse instrument. The argument for this choice of instrument would be that longer waiting times at customs may motivate foreign firms to source more inputs from the domestic market. The IV 2SLS regressions show that the interaction term with *inp* has a positive coefficient sign (column B, rows 3a and 3b), while it is negative in the OLS regression (column B, row 3a). Both interaction terms are insignificant. The interaction term with sectoral FDI technology intensity in a country (*tech*) has a negative but insignificant impact in both regressions (columns A and B, row 4), rejecting the findings of a positive impact (see, for example, Buckley, Wang, and Clegga 2007; Lin, Liub, and Zhanga 2009).

Unfortunately, many results fall short of the thresholds of statistical significance, which could be related to heterogeneous productivity levels of domestic firms (Girma and Görg 2007). We therefore examine the dynamics of productivity at different points of the distribution in a next step.

The Role of Domestic Firm Productivity
We split our sample into three productivity groups: $LP \leq 33.33$ percentile (low-productivity firms); 33.33 percentile $< LP \leq 66.67$ percentile (medium-productivity firms); and $LP > 66.67$ percentile (high-productivity firms). The IV 2SLS regression results by productivity level are shown in columns C to E. The FDI spillover variable (not shown in table 3.1) mostly shows negative coefficient signs for all types of firms, but is rarely significant. We detect clear differences

across the three types of domestic firms. Capital intensity (not shown in table 3.1) shows a constantly positive effect on LP, but the impact is strongest for high-productivity firms, followed by low-productivity firms and medium-productivity firms.

The interaction term with foreign ownership share (*own*) is insignificant for all types of firms (columns C to E, row 1). The interaction term with FDI sales to the domestic market (*market*) is positive and significant for both low- and medium-productivity firms, but more so for low-productivity firms (columns C and D, rows 2a and 2b). Although market-seeking FDI provides a higher spillover potential, the underlying transmission channels seem to be different for both types of firms. It is more likely that medium-productivity firms benefit from competition and demonstration effects, while low-productivity firms may rather benefit from potential integration in supply chains and labor turnover.

A sector's average percentage of domestic input purchases of FDI firms in a country (*inp*) also shows a positive and significant impact for low- and medium-productivity firms (columns C and D, row 3a). Our results suggest that more local sourcing is associated with more positive FDI spillovers to domestic firms. Various transmission channels between foreign firms and domestic suppliers are thinkable, for example, the demand and assistance effects, diffusion effect, and availability and quality effects (see chapter 2 for more details). Again, the effects are stronger for low-productivity firms, which may be the result of their higher catching-up potential.

The interaction term with foreign technology intensity (*tech*) is negative, but insignificant for all types of firms (columns C to E, row 4). However, the p value is lower for high-productivity firms, implying that a smaller technology gap between high-productivity domestic firms and foreign firms renders it more difficult for the former to absorb foreign technology.

In sum, the results show that low- and medium-productivity firms benefit more strongly from foreign presence than high-productivity firms in terms of both significance and coefficient sign.

The Role of Foreign Ownership Structure

In a next step, we also assess whether the results depend on the structure of foreign ownership, as discussed in chapter 2. We therefore calculated two variants of the FDI spillover variable in equation (3.2): (a) FDI spillovers based on firms with full foreign ownership, defined as a foreign participation of 100 percent, and (b) FDI spillovers based on firms with partial foreign ownership, defined as foreign participation of at least 10 percent but less than 100 percent. Annex 3B shows the number of foreign firms by structure of foreign ownership.

Analogously, we also calculated the average foreign firm characteristics in a sector and country, as described previously in the section "Measures of foreign spillover potential," based on fully owned and partially owned foreign firms only. Column F shows the results for full foreign ownership, while column G focuses on the results for partial foreign ownership. Although foreign firms in a country's sector sell on average 67.5 percent to the domestic market (*market*), this share is 69 percent for partial FDI versus 65 percent for fully owned FDI. Similarly, average purchases from local sources in a country's sector (*inp*) are 57 percent for all foreign firms, but the split sample shows it is almost 60 percent for partially owned FDI and only 55 percent for fully owned FDI (see summary statistics in annex 3C). These summary statistics alone suggest that partially foreign-owned firms are more integrated into domestic markets than fully foreign-owned firms, and may indicate a higher spillover potential for partially foreign-owned firms.

Note that the model dropped the sectoral foreign ownership variable (*own*) for fully foreign-owned firms, as the value, by definition, is 100 percent for every sector (column F, row 1). The interaction term with foreign firms' sales to the domestic market (*market*) is positive but only significant for fully foreign-owned FDI (column F, row 2b), implying that spillovers from market-oriented FDI are more beneficial from firms with full foreign ownership, possibly due to their higher willingness to transfer technology (Dimelis and Louri 2002; Takii 2005).

The interaction term with foreign firms' share of local inputs (*inp*) is positive and significant for partially foreign-owned FDI (column G, row 3a), suggesting that local sourcing increases the productivity spillovers from foreign firms with local participation, possibly due to the existence of more vertical linkages (Javorcik and Spatareanu 2008). On the other hand, more local sourcing reduces the spillovers from fully foreign-owned firms, which is insignificant.

Overall, the results cannot give a clear answer whether full or partial foreign ownership translates into higher FDI spillovers, since the latter also depend on the foreign firm's FDI motive and sourcing strategy.

Absorptive Capacity
Overall Results

Table 3.2 shows the regression results for domestic firms' absorptive capacities. Again, we only report the results for the interaction terms (for a full set of regression results, see Farole and Winkler 2012). Column A reports the estimates using OLS, while column B shows the results using IV 2SLS. FDI spillovers show a negative and often significant impact on *LP*, except for row (1). Interestingly, the results clearly indicate that the FDI spillover variable in the OLS estimates is upward biased compared to the IV 2SLS estimates (not shown in table 3.2). This confirms the results by Aitken and Harrison (1999), who find that foreign firms might be attracted to more-productive industries, which OLS estimates using cross-sectional data cannot adequately control for. Capital intensity (not shown in table 3.2) has a positive and significant effect in all specifications.

Absorptive capacities are measured at the domestic firm level and interacted with our FDI measure. The results show that absorptive capacity influences the role of FDI spillovers, regardless of the type of estimator being used. In particular, the interaction term with productivity gap (*gap*) shows a significantly positive impact (columns A and B, row 1).[20] This argues against the findings of other

Table 3.2 Absorptive Capacity, FDI Spillovers

		A	B	C	D	E	F	G
	Estimator:	OLS	IV 2SLS	IV 2SLS	IV 2SLS	IV 2SLS	IV 2SLS	IV 2SLS
Interaction term with FDI spillovers	Productivity level:	All	All	Low	Medium	High	All	All
	Foreign ownership:	All	All	All	All	All	Full	Partial
(1) gap_{isct}		1.6451***	1.7388***	1.6787***	0.7863***	1.5133***	2.1221***	2.8656***
		(0.000)	(0.000)	(0.000)	(0.000)	(0.000)	(0.000)	(0.000)
(2) $tech_{irst}$		0.4660***	0.4674***	0.1749**	0.0702	0.0869	0.4432***	0.8403***
		(0.000)	(0.000)	(0.013)	(0.101)	(0.148)	(0.000)	(0.000)
(3) $skills_{irst}$		−0.0178	0.1205	0.4030	0.1375	0.0655	−0.1315	0.6816
		(0.840)	(0.425)	(0.143)	(0.375)	(0.732)	(0.626)	(0.131)
(4) $size_{irst}$		0.2758***	0.3184***	0.1814**	0.0625	−0.0109	0.2635**	0.7147***
		(0.000)	(0.000)	(0.047)	(0.299)	(0.865)	(0.011)	(0.000)
(5) $aggl_{rct}$		0.5346**	1.0066*	2.3566***	0.8166	0.3288	−0.0537	1.8956***
		(0.046)	(0.054)	(0.001)	(0.178)	(0.702)	(0.950)	(0.008)
(6) exp_{irst}		0.6344***	0.9274***	0.0457	−0.1106	0.1904	0.9621***	1.8191***
		(0.000)	(0.000)	(0.841)	(0.388)	(0.286)	(0.001)	(0.000)

Source: Farole and Winkler 2012.
Note: FDI = foreign direct investment. Each column-row combination designates a separate regression. The dependent variable is labor productivity $lnlp_{irst}$. All regressions control for the foreign ownership share in a sector and country and capital intensity at the firm level and include sector, subnational region, and year fixed effects (see equation 3.3). Standard errors are clustered at the country-sector level. Low productivity: $LP \leq 33.33$ percentile. Medium productivity: 33.33 percentile $< LP \leq 66.67$ percentile. High productivity: $LP > 66.67$ percentile. Full foreign ownership: foreign participation = 100 percent. Partial foreign ownership: 10 percent ≥ foreign participation < 100 percent.
*$p < 0.1$, **$p < 0.05$, ***$p < 0.01$ (p values in parentheses). For the full set of regression results, see Farole and Winkler (2012).

studies claiming that a large technology gap is beneficial for local firms, since their catching-up potential increases (Findlay 1978; Wang and Blomström 1992; Smeets 2008), and instead supports the idea that too large a gap hinders absorption potential (Blalock and Gertler 2009).

The interaction with a domestic firm's technology (*tech*) has a significantly positive effect on domestic LP (columns A and B, row 2). This confirms the high number of studies pointing toward a positive effect of R&D (see chapter 2). The interaction term with skills (*skills*), however, does not influence domestic LP (columns A and B, row 3). Firm size, measured by number of employees (*size*), shows a positive interaction with FDI spillovers, which is significant (columns A and B, row 4). Our results confirm the majority of studies, which positively relate firm size to a domestic firm's capacity to absorb FDI spillovers.

The proximity to other firms in a subnational region (*aggl*) also shows a positive productivity effect in interaction with FDI spillovers (columns A and B, row 5), which is in line with the findings of Barrios, Luisito, and Strobl (2006). Finally, the interaction with exports, measured as export share (*exp*), also shows a significantly positive productivity effect (columns A and B, row 6), which is in line with other empirical studies (for example, Barrios and Strobl 2002; Schoors and van der Tol 2002; Lin, Liub, and Zhanga 2009; Jordaan 2011b) supporting the hypothesis that exporting renders domestic firms more competitive against negative rivalry effects created by multinationals (Crespo and Fontoura 2007).

Overall, the coefficients of the interaction terms are larger for the IV 2SLS estimates compared to the OLS regressions. In other words, absorptive capacities seem to have a bigger impact on FDI spillovers when controlling for the potential endogeneity between the FDI spillover variable and domestic firm productivity.

The Role of Domestic Firm Productivity

In a next step, we evaluate the role of domestic firm-level productivity for its absorptive capacity. Columns C to E show the IV 2SLS regression results for domestic firms by productivity level. FDI spillovers (not shown in table 3.2) are negative in most specifications across all firms, but only rarely significant. Capital intensity (not shown) again is significant and positive across all types of firms. The positive interaction term with technology gap (*gap*) is most important for firms with a low productivity level, almost equally high for high-productivity firms, whereas firms with a medium productivity level benefit less from a lower technology gap (columns C to E, row 1). The positive interaction terms with technology (*tech*) are only significant for low-productivity firms (column C, row 2), while medium- and high-productivity firms fall short of the 10 percent significance level (columns D and E, row 2). The interaction term with skills (*skills*) is also positive and narrowly misses the 10 percent significance level for low-productivity firms (column C, row 3).

The interaction term with firm size (*size*) is positive and significant for low-productivity firms (column C, row 4). Similarly, agglomeration (*aggl*) shows a significant and positive effect for low-productivity firms only (column C, row 5). This may reflect firm size and the fact that smaller firms tend to benefit more from urbanization externalities, while larger firms are more self-sufficient (Jacobs 1969). While the interaction terms with export share (*exp*) are insignificant across all types of firms (columns C to E, row 6), the *p* values and coefficient sizes reveal that export behavior matters more for highly productive firms, confirming that exporters are more productive (Bernard and Jensen 1999). Note that the effects on all domestic firms were significant (columns A and B, row 6).

In sum, the results tend to suggest that absorptive capacities are more important for low-productivity firms than for medium- and high-productivity firms, in terms of both statistical significance and in most cases also coefficient size.

The Role of Foreign Ownership Structure

In this section, we examine if foreign ownership structure influences the role of absorptive capacities of domestic firms in facilitating spillovers. Column F shows the IV 2SLS regression results for FDI from fully owned firms, while column G reports the results for partially owned firms. Interestingly, the FDI spillovers effect tends to be negative and significant for partial foreign ownership, while it often shows positive coefficient signs for full foreign ownership that are mostly insignificant (not shown in table 3.2). Capital intensity shows a positive and significant effect across both types of foreign ownership structure (not shown in table 3.2).

The results indicate clearly that absorptive capacities more strongly influence spillovers from FDI with local participation (partial FDI). A domestic firm's lower technology gap (*gap*, columns F and G, row 1), technology level (*tech*, columns F and G, row 2), size (*size*, columns F and G, row 4), and export share (*exp*, columns F and G, row 6) have a significantly stronger beneficial impact on productivity spillovers from partial FDI compared to full FDI. Proximity to other firms (*aggl*) shows a significantly positive productivity effect from partial FDI, whereas there is no effect from agglomeration in the presence of fully foreign-owned firms (columns F and G, row 5).

The findings suggest that FDI that includes partial domestic ownership participation better facilitates firms with absorptive capacity to reap the benefits of productivity spillovers, perhaps because partial-FDI firms offer greater opportunities for technology diffusion.

National Characteristics and Institutions
Overall Results

Finally, we focus on the role of host country characteristics and institutions in table 3.3 and report the results for the interaction terms only (for a full set of regression results, see Farole and Winkler 2012). The OLS regression results are shown in column A, while the IV 2SLS estimates are presented in column B. FDI spillovers mostly have a negative effect on LP that is rarely significant, while capital intensity shows a consistently positive impact (not shown in table 3.3). The interaction of our FDI measure with a country's government spending on education as percentage of GDP (*educ1*) has a strongly positive and significant productivity effect (columns A and B, row 3) in the OLS and IV 2SLS regressions, confirming the positive role of the local innovation infrastructure for FDI spillovers (see, for example, Tytell and Yudaeva 2007; Meyer and Sinani 2009).

Both trade indicators, namely a country's export share (*trade1*), and the absence of trade protectionism (*trade2*), are significant and interact positively with FDI spillovers (column A, rows 7 and 8). This supports the view that FDI spillovers are larger in countries that are more open toward trade (Meyer and Sinani 2009; Havranek and Irsova 2011; Du, Harrison, and Jefferson 2011). However, the interaction terms with the trade indicators are only significant in the OLS regressions.

While all other variables have no impact in the overall sample, the effect varies depending on the level of productivity of domestic firms as well as on the foreign ownership structure, as we show in the next two subsections.

The Role of Domestic Firm Productivity

Columns C to E show the IV 2SLS regression results by type of domestic firm productivity. The impact of FDI spillovers is insignificant across all specifications

Table 3.3 National Characteristics, FDI Spillovers

		A	B	C	D	E	F	G
	Estimator:	OLS	IV 2SLS	IV 2SLS	IV 2SLS	IV 2SLS	IV 2SLS	IV 2SLS
	Productivity level:	All	All	Low	Medium	High	All	All
	Foreign ownership:	All	All	All	All	All	Full	Partial
Interaction term with FDI spillovers								
(1)	$labor_{ct}$	0.0087	−0.8772	−0.2702	−1.1407	−0.4435	−2.5070	4.7343
		(0.979)	(0.465)	(0.874)	(0.479)	(0.797)	(0.118)	(0.117)
(2)	$finance_{ct}$	0.4316	−0.8907	0.1801	−0.7939	−0.2961	−2.8224	7.7047*
		(0.171)	(0.585)	(0.931)	(0.713)	(0.896)	(0.285)	(0.065)
(3)	$educ1_{ct}$	3.9472*	12.3485*	24.7543*	11.9357	6.0911	8.8373	35.4614**
		(0.069)	(0.087)	(0.078)	(0.110)	(0.466)	(0.567)	(0.041)
(4)	$educ2_{ct}$	−0.1922	−0.6653	0.1000	−1.1456	−1.6284	−2.0013*	2.6142
		(0.778)	(0.544)	(0.953)	(0.381)	(0.314)	(0.099)	(0.257)
(5)	rd_{ct}	−35.1017	−53.2774	−141.2323	−126.9857	−135.8625	−216.5904	367.1298
		(0.177)	(0.691)	(0.549)	(0.541)	(0.548)	(0.442)	(0.655)
(6)	$investm_{ct}$	−0.2561	−0.0952	0.9992	−0.0908	−0.5215	−1.2063	3.3987**
		(0.274)	(0.913)	(0.429)	(0.930)	(0.662)	(0.120)	(0.029)
(7)	$trade1_{ct}$	0.5713**	2.6208	7.2220**	2.9364	0.4347	−3.5995	10.3784***
		(0.038)	(0.205)	(0.026)	(0.196)	(0.890)	(0.514)	(0.004)
(8)	$trade2_{ct}$	0.6309**	0.6313	3.2760	0.5712	−1.2510	−2.1743	4.1490*
		(0.019)	(0.664)	(0.133)	(0.720)	(0.527)	(0.220)	(0.070)
(9)	$business_{ct}$	−0.3019	1.4287	3.2913	−0.7000	−0.0656	−0.9114	−4.7106
		(0.484)	(0.642)	(0.560)	(0.808)	(0.987)	(0.791)	(0.381)
(10)	hhi_{sct}	0.2488	0.5884	1.5837**	0.5752	−0.2433	0.3176	0.8914
		(0.316)	(0.293)	(0.038)	(0.417)	(0.793)	(0.708)	(0.276)
(11)	$income_{ct}$	0.0113	0.0468	0.2871	0.0298	−0.1582	−0.2099	0.5708**
		(0.873)	(0.736)	(0.196)	(0.841)	(0.402)	(0.216)	(0.036)

Source: Farole and Winkler 2012.

Note: FDI = foreign direct investment. Each column-row combination designates a separate regression. The dependent variable is labor productivity $\ln lp_{ifst}$. All regressions control for the foreign ownership share in a sector and country and capital intensity at the firm level and include sector, subnational region, and year fixed effects (see equation 3.3). Standard errors are clustered at the country-sector level. Low productivity: $LP \leq 33.33$ percentile. Medium productivity: 33.33 percentile $< LP \leq 66.67$ percentile. High productivity: $LP > 66.67$ percentile. Full foreign ownership: foreign participation = 100 percent. Partial foreign ownership: 10 percent $\geq$ foreign participation $<$ 100 percent. *$p < 0.1$, **$p < 0.05$, ***$p < 0.01$ (p values in parentheses). For the full set of regression results, see Farole and Winkler (2012).

(not shown in table 3.3). The positive impact of capital intensity matters most for high- and low-productivity firms, followed by medium-productivity firms (not shown in table 3.3).

The interaction term with a country's government spending on education as percentage of GDP (*educ1*) has a significantly positive productivity effect only for low-productivity firms (column C, row 3). It is also interesting to take a look at the coefficient sizes and p values across the three types of firms, suggesting that skilled labor becomes more important the less productive domestic firms are.

A country's export share (*trade1*) positively interacts with FDI spillovers for low-productivity firms (column C, row 7). While the effect is statistically insignificant for medium- and high-productivity firms, the coefficient sizes and p values across the three types of firms suggest that an open trade regime is less important for more productive firms. The interaction term with the absence of trade protectionism (or openness to imports) (*trade2*) also shows a positive effect for low-productivity firms, but narrowly misses the 10 percent significance level (column C, row 8).

Finally, the interaction term with sector concentration (*hhi*) shows a positive and significant impact on LP for firms with low productivity levels (column C, row 10). Our findings confirm the results of Barrios and Strobl (2002) showing that higher concentration of sectoral activity has positive productivity effects that are stronger for low-productivity firms. The interaction term with income per capita (*income*) is positive, but falls short of the threshold levels of statistical significance for low-productivity firms (column C, row 11).

In sum, the interaction terms are only significant for low-productivity firms, confirming the previous finding that MFs matter more for lower-productivity firms. Among the various institutional variables we test, only a country's education spending, trade openness, and sector concentration seem to mediate the spillover effects from FDI.

The Role of Foreign Ownership Structure
Finally, we focus on the role of foreign ownership structure for national characteristics. Columns F and G report the IV 2SLS regression results for FDI spillovers from fully owned versus partially owned firms. The results show that national MFs influence FDI spillovers differently, depending on the type of foreign ownership.

More flexible labor markets (*labor*) positively affect them from partially foreign-owned firms, but negatively influence productivity spillovers from fully foreign-owned firms (columns F and G, row 1). Both effects, however, fall short of the threshold levels of statistical significance. Fully foreign-owned firms might be more reluctant to transfer technology in a flexible labor market due to the risk of technology diffusion. Foreign firms with domestic participation, in contrast, tend to be characterized by more supply linkages with the host country. In a flexible labor market, this could be beneficial to knowledge and productivity spillovers due to higher labor turnover.

A country's financial freedom (*finance*) positively and significantly mediates the productivity effect from partially foreign-owned firms, but has no effect on the productivity effect generated by fully foreign-owned firms (columns F and G, row 2). It is possible that partially foreign-owned firms bring in scarce capital to developing countries (Harrison, Love, and McMillan 2004), which seems to be facilitated by higher financial freedom. This would ease access to finance for local firms and strengthen their absorptive capacities.

Government spending on education (*educ1*) significantly increases the productivity effect from partial foreign ownership, while there is no effect from fully foreign-owned firms (columns F and G, row 3). This supports earlier findings that absorptive capacities contribute to greater productivity spillovers in the presence of partially foreign-owned firms than fully foreign-owned firms. A higher share of people with completed secondary and tertiary education as percentage of population (*educ2*) also interacts positively with spillovers from partially foreign-owned firms, although the effect is not statistically significant (column G, row 4). Both interaction effects suggest that economies with higher skill availability benefit more strongly from partially foreign-owned firms, possibly due to more labor turnover of skilled labor to domestic firms.

Interestingly, higher skill availability in a country (*educ2*) significantly lowers the spillovers from fully foreign-owned firms (column F, row 4). It may be possible that economies with a higher share of skilled people attract more fully foreign-owned firms whose higher technology intensity could crowd-out domestic producers, supporting the findings of Abraham, Konings, and Slootmaekers (2010). Such firms could also bid away high-quality labor from domestic firms, resulting in negative spillover effects.

A country's investment openness (*investm*) positively and significantly interacts with FDI spillovers from partially foreign-owned firms, while the effect from fully foreign-owned firms is negative but narrowly misses the 10 percent level of significance (columns F and G, row 6). More investment freedom could attract more partially foreign-owned firms and thus increase the spillover potential due to technology leakages and more linkages to the host country. At the same time, investment openness could also contribute to a greater number of "footloose" fully foreign-owned firms, with fewer linkages to the local economy and crowding-out potential.

Interestingly, a country's export share in GDP (*trade1*) significantly increases the spillover impact from partial foreign ownership, while the effect from full foreign ownership is negative but insignificant (columns F and G, row 7). Analogously, trade freedom (*trade2*)—reflecting the absence of trade protectionism including imports—positively moderates the productivity impact from partial foreign ownership, while the effect from fully foreign-owned firms is negative but again insignificant (columns F and G, row 8). Partially foreign-owned firms have a higher spillover potential because they are more integrated into their host countries in terms of sales and local sourcing (see annex 3C). Nevertheless, they may still depend on exports or imported inputs, hence, trade openness and freedom positively interacts with FDI spillovers from such firms. Fully foreign-owned firms, in contrast, are more likely to be export platforms with fewer local linkages and technology leakages.

Finally, income per capita (*income*) positively mediates the FDI spillover effect on domestic LP from partially foreign-owned firms, while there is no effect for fully foreign-owned firms (columns F and G, row 11). This reflects the importance of national competitiveness and development, given the possibility of more local linkages for partially foreign-owned firms.

Overall, the results clearly indicate that spillovers from partially foreign-owned firms are positively mediated by various national and institutional factors. In contrast, national and institutional characteristics exert no or even a negative effect on spillovers from fully foreign-owned firms.

Conclusions

In this chapter, we use a cross-section of more than 25,000 manufacturing firms in 78 LMICs from the World Bank's Enterprise Surveys Indicator Database to identify the MFs that determine intra-industry productivity spillovers. Besides OLS, this paper also applies an instrumental variables approach to addresses the potential endogeneity between FDI spillovers and domestic firm productivity. We differentiate between three major types of MFs, namely the FDI spillover potential, domestic firms' absorptive capacity, and differences in host country factors and institutions. This is the first study to our knowledge that interacts FDI with all three types of MFs against productivity outcomes.

We find evidence for negative FDI spillovers overall, which we relate to increased competition in the short term for market share (both domestic and export) and skilled labor. Regarding the FDI spillover potential, we find a positive and significant impact in the overall sample for market orientation in the OLS regressions—the higher the share of FDI output sold domestically, the greater productivity spillovers to domestic firms. Splitting the sample into three productivity groups, we find that larger output share sold to the domestic market and larger share of local inputs contributes to higher spillovers for low- and medium-productivity firms.

Regarding a domestic firm's absorptive capacity, almost all variables show the expected coefficient signs, suggesting that a lower technology gap, firm technology, firm size, proximity to other firms, and export behavior interact positively with FDI spillovers, although results vary by type of firm. In sum, the MFs are more significant for low-productivity firms, confirming that absorptive capacity correlates with firm-level productivity.

The overall results for national and institutional MFs suggest that a country's spending on education as well as trade openness (the latter can only be confirmed in the OLS regressions) matter most. However, again the results vary significantly depending on the productivity levels of domestic firms. The positive interaction with education spending and trade is only significant for low-productivity firms and tends to fall with higher firm-level productivity. Splitting the sample into three productivity levels reveals that a country's institutional framework only seems to benefit low-productivity firms. A country's spending on education, export share, and sector concentration significantly increase FDI spillovers.

In sum, all three types of MFs have an influence on domestic firms, although the actual effect is clearly related to domestic firm productivity. Low-productivity firms tend to benefit from FDI spillover potential and the institutional framework and also exhibit the highest absorptive capacity. The MFs might help them counterbalance some of the negative spillover effects from FDI, despite their productivity disadvantage. Medium-productivity firms seem to be disadvantaged in terms of the institutional framework and also show lower benefits from FDI spillover potential and absorptive capacity. High-productivity firms only benefit from absorptive capacities—in particular a low productivity gap—while the FDI spillover potential and the institutional framework do not or negatively influence FDI spillovers for those firms.

We also assess whether the results depend on the structure of foreign ownership. Across all three types of MFs, we find strong evidence that partial foreign ownership shows more positive FDI spillovers on domestic firm productivity. Regarding the FDI spillover potential, the spillovers from partially foreign-owned firms are larger the higher the share of local sourcing is, which does not hold for fully foreign-owned firms. On the other hand, foreign firms' sales to the domestic market exert a positive effect on FDI spillovers, which is only significant for fully foreign-owned firms.

Our results also indicate that a domestic firm's absorptive capacities—in particular a smaller technology gap, higher technology level, larger size, more agglomeration, and higher export share—more strongly influence spillovers from FDI in the presence of FDI that includes local partners (partial foreign ownership) than they do in the presence of fully foreign-owned firms.

Finally, the role of local participation for the moderating effects of national characteristics is clearly positive. FDI from partially foreign-owned firms positively interacts with a country's financial freedom, government spending on education, investment freedom, export share, freedom from trade protectionism, and income level. In contrast, national and institutional characteristics exert no or even a negative effect on spillovers from fully foreign-owned firms.

Overall, these findings suggest that policies designed to promote greater spillovers from foreign investment need to take into account not only these three important MFs—FDI spillover potential, absorptive capacity of domestic firms, and the host country environment—but also, critically, the heterogeneity of firms in the host country. The effectiveness of interventions is likely to vary significantly depending on the characteristics of domestic firms, as well as the form of foreign investment.

Annex 3A Survey Year and Number of Domestic and Foreign Manufacturing Firms by Country

Country	Survey year	Domestic manufacturing firms	Foreign manufacturing firms	Country	Survey year	Domestic manufacturing firms	Foreign manufacturing firms
Albania	2007	87	23	Mauritius	2009	157	18
Algeria	2007	374	10	Mexico	2010	1,046	110
Argentina	2010	681	111	Moldova	2009	86	22
Armenia	2009	102	12	Mongolia	2009	115	15
Azerbaijan	2009	95	23	Morocco	2007	354	103
Belarus	2008	86	13	Mozambique	2007	284	56
Bosnia and Herzegovina	2009	113	10	Namibia	2006	114	37
Bolivia	2010	118	22	Nepal	2009	124	4
Brazil	2009	851	59	Nicaragua	2010	115	10
Burkina Faso	2009	81	16	Nigeria	2007	938	10
Burundi	2006	110	28	Pakistan	2007	763	20
Cameroon	2009	87	29	Panama	2010	101	14
Chile	2010	673	102	Paraguay	2010	108	17
Colombia	2010	641	65	Peru	2010	673	87
Costa Rica	2010	272	53	Philippines	2009	692	262
Côte d'Ivoire	2009	144	49	Poland	2009	137	14
Croatia	2007	364	47	Romania	2009	143	36
Ecuador	2010	97	22	Russian Federation	2009	660	39
Egypt, Arab Rep.	2008	1,103	36	Rwanda	2006	54	14
El Salvador	2010	104	20	Senegal	2007	243	16
Ethiopia	2006	337	22	Serbia	2009	117	19
Georgia	2008	109	14	South Africa	2007	579	101
Ghana	2007	271	19	St. Vincent and the Grenadines	2010	124	24
Guatemala	2010	315	40	Swaziland	2006	68	38
Guinea	2006	122	15	Tajikistan	2008	98	15
Guinea-Bissau	2006	72	9	Tanzania	2006	247	39
Honduras	2010	130	18	Thailand	2006	800	230
India	2006	2,134	37	Turkey	2008	866	29
Indonesia	2009	1,067	89	Uganda	2006	276	58
Jamaica	2010	103	16	Ukraine	2008	523	46
Jordan	2006	305	48	Uruguay	2010	327	32
Kazakhstan	2009	169	14	Uzbekistan	2008	87	34
Kenya	2007	331	65	Venezuela, RB	2010	71	14
Kyrgyz Republic	2009	80	16	Vietnam	2009	649	130
Latvia	2009	65	24	Yemen, Rep.	2010	238	5
Lithuania	2009	85	19	Yugoslavia, former	2009	98	23
Madagascar	2009	125	79	Zambia	2007	236	68
Malaysia	2007	773	321	SUM		25,199	3,440
Mauritania	2006	112	16				

Source: Enterprise Surveys Indicator Database.

Annex 3B Number of Domestic and Foreign Manufacturing Firms by Sector

| | Domestic firms | | Foreign firms | | | | | |
| | All | | All | | Full foreign ownership | | Partial foreign ownership | |
Sector	No.	%	No.	%	No.	%	No.	%
Manufacturing	25,199	100.0	3,440	100.0	1,789	100.0	1,651	100.0
Textiles	2,051	8.1	194	5.6	106	5.9	88	5.3
Leather	275	1.1	42	1.2	25	1.4	17	1.0
Garments	3,439	13.6	373	10.8	217	12.1	156	9.4
Food	5,098	20.2	632	18.4	309	17.3	323	19.6
Metals and machinery	3,878	15.4	534	15.5	286	16.0	248	15.0
Electronics	697	2.8	188	5.5	103	5.8	85	5.1
Chemicals and pharmaceuticals	1,986	7.9	391	11.4	198	11.1	193	11.7
Wood and furniture	715	2.8	46	1.3	16	0.9	30	1.8
Nonmetallic and plastic materials	2,653	10.5	446	13.0	218	12.2	228	13.8
Auto and auto components	370	1.5	75	2.2	37	2.1	38	2.3
Other manufacturing	4,037	16.0	519	15.1	274	15.3	245	14.8

Source: Enterprise Surveys Indicator Database.
Note: Full foreign ownership: foreign participation = 100 percent. Partial foreign ownership: 10 percent ≥ foreign participation < 100 percent.

Annex 3C Summary Statistics

Variable	Observation	Mean	Standard deviation	Minimum	Maximum
$lnlp_{irst}$	18,287	8.4083	1.7609	−11.4958	20.4382
FDI^y_{sct}	25,660	0.2648	0.2381	0.0000	1.0000
$FDI^y_{sct,full}$	25,660	0.1406	0.1863	0.0000	0.9927
$FDI^y_{sct,partial}$	25,660	0.1243	0.1556	0.0000	1.0000
$lncapint_{irst}$	18,807	7.7379	2.1370	−4.0956	19.7083
FDI spillover potential					
own_{sct}	24,433	0.7251	0.1865	0.1000	1.0000
$own_{sct,full}$	21,002	1.0000	0.0000	1.0000	1.0000
$own_{sc,partial}$	22,995	0.5232	0.1731	0.1000	0.9900
$market_{sct}$	24,433	0.6745	0.2695	0.0000	1.0000
$market_{sc,full}$	20,720	0.6480	0.3118	0.0000	1.0000
$market_{sct,partial}$	22,923	0.6935	0.2704	0.0000	1.0000
inp_{sct}	24,330	0.5695	0.2393	0.0000	1.0000
$inp_{sct,full}$	20,680	0.5506	0.2892	0.0000	1.0000
$inp_{sct,partial}$	22,500	0.5968	0.2455	0.0000	1.0000
$tech_{sct}$	24,433	2.4402	0.7434	0.0000	4.0000
$tech_{sct,full}$	20,863	2.3593	0.8844	0.0000	4.0000
$tech_{sct,partial}$	22,995	2.5152	0.7710	0.0000	4.0000

table continues next page

Annex 3C Summary Statistics *(continued)*

Variable	Observation	Mean	Standard deviation	Minimum	Maximum
Absorptive capacity					
gap_{isct}	16,145	−0.6668	1.5280	−19.0525	7.4185
$tech_{irst}$	29,770	1.3461	1.1575	0.0000	4.0000
$skills_{irst}$	24,782	0.6381	0.3283	0.0000	1.0000
$size_{irst}$	29,383	3.5342	1.3551	0.0000	8.6995
$agglr_{ct}$	29,770	0.3451	0.2797	0.0017	1.0000
exp_{irst}	29,471	0.1278	0.2825	0.0000	1.0000
National characteristics and institutional framework					
$labor_{ct}$	28,992	4.0926	0.2236	3.2619	4.6052
$finance_{ct}$	28,992	3.7619	0.3497	2.3026	4.3820
$educ1_{ct}$	21,091	0.1701	0.0424	0.0441	0.2746
$educ2_{ct}$	26,205	0.2579	0.1445	0.0120	0.7200
rd_{ct}	16,743	0.0058	0.0031	0.0003	0.0110
$investment_{ct}$	28,992	3.7998	0.3424	1.6094	4.4427
$trade1_{ct}$	28,830	0.3361	0.1963	0.0453	1.1002
$trade2_{ct}$	28,992	4.1718	0.3322	3.1781	4.4773
$business_{ct}$	28,992	4.0884	0.2022	2.9069	4.4998
hhi_{sct}	25,804	0.1526	0.1587	0.0153	1.0000
$income_{ct}$	29,238	7.3396	1.1107	4.5344	9.6946
Instruments					
$FDI^{Y}_{sc,lag}$	17,962	0.2730	0.2391	0.0000	0.9928
$FDI^{Y}_{sc,lag,full}$	17,962	0.1424	0.1872	0.0000	0.9927
$FDI^{Y}_{sc,lag,partial}$	17,962	0.1306	0.1615	0.0000	0.9740
$lnipa\_length_{sc,2000-2004}$	12,839	0.4321	0.7122	0.0000	1.7918
$lntariff_{sct}$	25,808	2.0668	0.8026	0.0000	4.2277
$lnexp\_customs_{sct}$	24,863	1.6805	0.7733	0.0000	5.1930
$lnimp\_customs_{sct}$	25,456	2.2423	0.6483	0.0000	4.1972

Note: FDI = foreign direct investment.

Notes

1. The authors would like to thank Holger Görg, Beata Javorcik, and Ben Shepherd for valuable comments on a previous version of this paper and Tofinn Harding, Daniel Lederman, Minh Cong Nguyen, Federica Saliola, and Murat Seker for help with the data.
2. A more detailed version of this chapter is published in Farole and Winkler (2012).
3. In addition, their study also differs in terms of methodology, as it uses the *t* statistic of foreign direct investment (FDI) spillover estimates from existing research as the dependent variable.
4. Blalock and Gertler (2009) calculate FDI spillovers at the region-sector level due to the geographic specificities of Indonesia (that is, many islands). We follow the majority of studies that calculate FDI at the sectoral level.

5. Our cross-sectional dataset is not suited to calculate total factor productivity using the standard methodologies that control for the endogeneity of input demand. See, for example, Olley and Pakes (1996) or Levinsohn and Petrin (2003).
6. Capital stock is defined as the replacement value of machinery, vehicles, and equipment. Information on the replacement value of buildings and land was very incomplete and therefore was not included.
7. See http://www.enterprisesurveys.org/~/media/FPDKM/EnterpriseSurveys/Documents/Misc/Indicator-Descriptions.pdf for a description of the indicators. Our analysis is based on the October 2011 release of the Enterprise Surveys Indicator Database.
8. We thank Federica Saliola and Murat Seker for making these data available to us.
9. http://data.worldbank.org/data-catalog/world-development-indicators.
10. Only Kosovo did not fulfill this criterion.
11. We drop these, as the database only included 11 high-income countries, which were not representative of high-income countries (some Organisation for Economic Co-operation and Development [OECD] countries, some non-OECD countries, and one oil exporter).
12. We thank Minh Cong Nguyen for making an older version of the Enterprise Surveys Indicator Database available to us.
13. The authors are grateful to Torfinn Harding, Beata Javorcik, and Daniel Lederman for making this dataset available to us.
14. http://wits.worldbank.org/wits/.
15. http://www.wto.org/english/tratop_e/dda_e/status_e/nama_e.htm.
16. We included services and foreign firms in this measure as urbanization economies refer to spillovers due to the proximity of firms from all types of sectors and ownership.
17. http://data.worldbank.org/data-catalog/world-development-indicators.
18. The number of observations is lower in the IV 2SLS regressions since not all instruments are available for all countries (see summary statistics in annex 3C).
19. Görg and Strobl (2001) and Görg and Greenaway (2004), however, argue that studies using cross-sectional data tend to find positive FDI spillovers, which they relate to potential time-invariant effects across units (firms or sectors) that are correlated with the FDI spillover variable without being caused by it (endogeneity). For example, foreign firms may invest more strongly in more productive sectors, which would result in a positive association between sectoral FDI and productivity, although we would have a case of reverse causality. However, as this example shows, endogeneity tends to be a bigger problem for sector-level studies than for firm-level studies, especially if the latter control for sector-fixed effects.
20. Recall that our variable *gap* measures the ratio of domestic firm productivity to the median foreign firm productivity, so a higher number indicates a lower gap.

References

Abraham, F., J. Konings, and V. Slootmaekers. 2010. "FDI Spillovers in the Chinese Manufacturing Sector: Evidence of Firm Heterogeneity." *Economies of Transition* 18 (1): 143–82.

Aitken, B., and A. Harrison. 1999. "Do Domestic Firms Benefit from Direct Foreign Investment? Evidence from Venezuela." *The American Economic Review* 89 (3): 605–18.

Barrios, S., B. Luisito, and E. Strobl. 2006. "Coagglomeration and Spillovers." *Regional Science and Urban Economics* 36 (4): 467–81.

Barrios, S., and E. Strobl. 2002. "Foreign Direct Investment and Productivity Spillovers: Evidence from the Spanish Experience." *Review of World Economics* 138 (3): 459–81.

Barro, R., and J.W. Lee. 2010. "A New Data Set of Educational Attainment in the World, 1950–2010." NBER Working Paper 15902, National Bureau of Economic Research (NBER), Cambridge, MA.

Bernard, A., and J. Jensen. 1999. "Exceptional Exporter Performance: Cause, Effect or Both?" *Journal of International Economics* 47 (1): 1–25.

Blalock, G., and P. Gertler. 2009. "How Firm Capabilities Affect Who Benefits from Foreign Technology." *Journal of Development Economics* 90 (2): 192–99.

Buckley, P., C. Wang, and J. Clegga. 2007. "The Impact of Foreign Ownership, Local Ownership and Industry Characteristics on Spillover Benefits from Foreign Direct Investment in China." *International Business Review* 16 (2): 142–58.

Castellani, D., and A. Zanfei. 2003. "Technology Gaps, Absorptive Capacity and the Impact of Inward Investments on Productivity of European Firms." *Economics of Innovation and New Technology* 12 (6): 555–76.

Crespo, N., and M. Fontoura. 2007. "Determinant Factors of FDI Spillovers—What Do We Really Know?" *World Development* 35 (3): 410–25.

Dimelis, S., and H. Louri. 2002. "Foreign Ownership and Production Efficiency: A Quantile Regression Analysis." *Oxford Economic Papers* 54 (3): 449–69.

Djankov, S., and B. Hoekman. 2000. "Foreign Investment and Productivity Growth in Czech Enterprises." *World Bank Economic Review* 14 (1): 49–64.

Du, L., A. Harrison, and G. Jefferson. 2011. "Do Institutions Matter for FDI Spillovers? The Implications of China's 'Special Characteristics'." NBER Working Paper 16767, National Bureau of Economic Research (NBER), Cambridge, MA.

Farole, T., and D. Winkler. 2012. "Foreign Firm Characteristics, Absorptive Capacity and the Institutional Framework: The Role of Mediating Factors for FDI Spillovers in Low- and Middle-Income Countries." Policy Research Working Paper 6265, World Bank, Washington, DC.

Findlay, R. 1978. "Relative Backwardness, Direct Foreign Investment, and the Transfer of Technology: A Simple Dynamic Model." *Quarterly Journal of Economics* 92 (1): 1–16.

Girma, S., and H. Görg. 2007. "The Role of Efficiency Gap for Spillovers from FDI: Evidence from the U.K. Electronics and Engineering Sectors." *Open Economies Review* 18 (2): 215–32.

Görg, H., and D. Greenaway. 2004. "Much Ado about Nothing? Do Domestic Firms Really Benefit from Foreign Direct Investment?" *The World Bank Research Observer* 19 (2): 171–97.

Görg, H., and E. Strobl. 2001. "Multinational Companies and Productivity Spillovers: A Meta-Analysis." *Economic Journal* 111 (475): F723–39.

Harding, T., and B. Javorcik. 2011. "Foreign Direct Investment and Export Upgrading." Economics Series Working Papers 526, University of Oxford, Department of Economics. Forthcoming in *The Review of Economics and Statistics*.

Harrison, A. 1994. "Productivity, Imperfect Competition and Trade Reform." *Journal of International Economics* 36 (1–2): 53–73.

Harrison, A., I. Love, and M. McMillan. 2004. "Global Capital Flows and Financing Constraints." *Journal of Development Economics* 75 (1): 269–301.

Haskel, J., S. Pereira, and M. Slaughter. 2007. "Does Inward Foreign Direct Investment Boost the Productivity of Domestic Firms?" *The Review of Economics and Statistics* 89 (3): 482–96.

Havranek, T., and Z. Irsova. 2011. "Estimating Vertical Spillovers from FDI: Why Results Vary and What the True Effect Is." *Journal of International Economics* 85 (2): 234–44.

Hoekman, B., and B. Javorcik, eds. 2006. "Lessons from Empirical Research on Technology Diffusion through Trade and Foreign Direct Investment." In *Global Integration and Technology Transfer*, 1–28. Washington, DC: World Bank; New York: Palgrave Macmillan.

Iacovone, L., B. Javorcik, W. Keller, and J. Tybout. 2011. "Supplier Responses to Wal-Mart's Invasion of Mexico." NBER Working Paper 17204, National Bureau of Economic Research (NBER), Cambridge, MA.

Jacobs, J. 1969. *The Economy of Cities.* New York: Random House.

Javorcik, B. 2004. "The Composition of Foreign Direct Investment and Protection of Intellectual Property Rights: Evidence from Transition Economies." *European Economic Review* 48 (1): 39–62.

Javorcik, B., and M. Spatareanu. 2008. "To Share or Not to Share: Does Local Participation Matter for Spillovers from Foreign Direct Investment?" *Journal of Development Economics* 85 (1–2): 194–221.

Jordaan, J. 2011a. "Cross-Sectional Estimation of FDI Spillovers When FDI is Endogenous: OLS and IV Estimates for Mexican Manufacturing Industries." *Applied Economics* 43 (19): 2451–63.

———. 2011b. "Local Sourcing and Technology Spillovers to Mexican Suppliers: How Important Are FDI and Supplier Characteristics?" *Growth and Change* 42 (3): 287–319.

Keller, W., and S. Yeaple. 2009. "Multinational Enterprises, International Trade, and Productivity Growth: Firm-Level Evidence from the United States." *The Review of Economics and Statistics* 91 (4): 821–31.

Konings, J. 2001. "The Effects of Foreign Direct Investment on Domestic Firms: Evidence from Firm-Level Panel Data in Emerging Economies." *Economics of Transition* 9 (3): 619–33.

Levinsohn, J., and A. Petrin. 2003. "Estimating Production Functions Using Inputs to Control for Unobservables." *Review of Economic Studies* 70 (2): 317–42.

Lin, P., Z. Liub, and Y. Zhanga. 2009. "Do Chinese Domestic Firms Benefit from FDI Inflow? Evidence of Horizontal and Vertical Spillovers." *China Economic Review* 20 (4): 677–91.

Lipsey, R., and F. Sjöholm. 2005. "The Impact of Inward FDI on Host Countries: Why Such Different Answers?" In *Does Foreign Direct Investment Promote Development?* edited by T. Moran, E. Graham, and M. Blomström, 23–43. Washington, DC: Institute for International Economics and Center for Global Development.

Melitz, M. 2003. "The Impact of Trade on Intra-Industry Reallocations and Aggregate Industry Productivity." *Econometrica* 71 (6): 1695–725.

Meyer, K., and E. Sinani. 2009. "When and Where Does Foreign Direct Investment Generate Positive Spillovers? A Meta-Analysis." *Journal of International Business Studies* 40 (7): 1075–94.

Milberg, W., and D. Winkler. 2013. *Outsourcing Economics: Global Value Chains in Capitalist Development.* New York: Cambridge University Press.

Olley, G., and A. Pakes. 1996. "The Dynamics of Productivity in the Telecommunications Equipment Industry." *Econometrica* 64 (6): 1263–97.

Paus, E., and K. Gallagher. 2008. "Missing Links: Foreign Investment and Industrial Development in Costa Rica and Mexico." *Studies of Comparative International Development* 43 (1): 53–80.

Schoors, K., and B. van der Tol. 2002. "Foreign Direct Investment Spillovers within and between Sectors: Evidence from Hungarian Data." Working Paper 2002/157, University of Gent.

Sinani, E., and K. Meyer. 2004., "Spillovers of Technology Transfer from FDI: The Case of Estonia." *Journal of Comparative Economics* 32 (3): 445–66.

Smeets, R. 2008. "Collecting the Pieces of the FDI Knowledge Spillovers Puzzle." *World Bank Research Observer* 23 (2): 107–38.

Takii, S. 2005. "Productivity Spillovers and Characteristics of Foreign Multinational Plants in Indonesian Manufacturing, 1990–1995." *Journal of Development Economics* 76 (2): 521–42.

Tytell, I., and K. Yudaeva. 2007. "The Role of FDI in Eastern Europe and New Independent States: New Channels for the Spillover Effect." In *Foreign Direct Investment in Europe: A Changing Landscape*, edited by K. Liebscher, J. Christl, P. Mooslechner, and D. Ritzberger-Grünwald, 76–86. Cheltenham, U.K.: Edward Elgar.

Wang, J., and M. Blomström. 1992. "Foreign Investment and Technology Transfer. A Simple Model." *European Economic Review* 36 (1): 137–55.

World Bank. 2011. "Enterprise Surveys Indicator Database." Enterprise Analysis Unit, World Bank, Washington, DC. http://www.enterprisesurveys.org/~/media/FPDKM/EnterpriseSurveys/Documents/Misc/Indicator-Descriptions.pdf.

CHAPTER 4

Determining the Nature and Extent of Spillovers: Empirical Assessment

Deborah Winkler[1,2]

Abstract

Using unique survey data collected as part of this research in Chile, Ghana, Kenya, Lesotho, Mozambique, Swaziland, and Vietnam, we first evaluate whether foreign investors differ from domestic investors in terms of their potential to generate positive spillovers for local suppliers. We find that while foreign investors might offer higher potential for spillovers due to stronger performance characteristics relative to domestic counterparts, this is mitigated by the fact that foreign investors have fewer linkages with the local economy and offer less supplier assistance. We also find that specific characteristics of foreign investors mediate both linkages and assistance extended to local suppliers.

Additionally, we examine the role of suppliers' absorptive capacities in determining the intensity of their linkages with multinationals. Our results indicate that several supplier characteristics matter, but these effects also depend on the length of the supplier relationship. Finally, we confirm the existence of positive effects of assistance (including technical audits, joint product development, and technology licensing) on foreign direct investment spillovers, while we find no evidence for demand effects.

Introduction

Using newly collected survey data on direct supplier-multinational linkages in Chile, Ghana, Kenya, Lesotho, Mozambique, Swaziland, and Vietnam, this chapter presents the findings of an empirical assessment of the nature and extent of spillovers. We first evaluate whether foreign investors differ from domestic producers in terms of their overall performance, linkages with the local economy, and supplier assistance, which all influence firms' potential to generate productivity spillovers. We also study the relationship between foreign investor characteristics and linkages with the local economy as well as assistance extended to local suppliers. We then shift the focus to domestic suppliers and examine the role of supplier firm characteristics—the so-called absorptive capacities—for their linkages with multinationals. Finally, focusing on assistance and demand effects, we assess how factors within the transmission channels between multinationals and local suppliers affect foreign direct investment (FDI) spillovers.

Studies on FDI spillovers that focus on direct supplier-multinational linkages based on foreign investor or supplier survey data are rare. Focusing on foreign affiliates in five transition economies, Giroud, Jindra, and Marek (2012) find that foreign firm characteristics have a positive impact on backward FDI linkages and spillovers. Javorcik and Spatareanu (2009) find evidence for "learning-by-supplying" for a sample of Czech manufacturing firms, although there is also evidence for self-selection into supplying due to higher productivity ex ante. Jordaan (2011) also confirms the existence of positive backward spillovers on manufacturing suppliers in Mexico. Specifically, positive spillovers are facilitated through supplier firms' absorptive capacities and the level of support from the multinational. Studying the Polish automotive sector, Gentile-Lüdecke and Giroud (2012) examine the mechanisms behind knowledge spillovers of suppliers. While the authors don't find evidence for a supporting role of suppliers' absorptive capacities on knowledge acquisition, they find evidence for a supportive role on performance improvement and new knowledge creation.

This chapter is structured as follows. The next section compares foreign investors and domestic producers in terms of their potential to generate productivity spillovers and also studies the role of foreign investor characteristics for their FDI spillover potential. The third section then evaluates the role of suppliers' absorptive capacities for FDI linkages, while the fourth section analyzes various factors within the transmission channels between suppliers and multinationals that increase FDI spillovers. Section five concludes.

Which Foreign Investor Characteristics Increase the FDI Spillover Potential?

This section focuses on the role of foreign investor characteristics for the FDI spillover potential. The first subsection presents the dataset being used in this section. The following subsection evaluates if there are differences between foreign investors and domestic producers in terms of their potential to generate positive spillovers. The third subsection examines if there are differences in the extent of FDI spillover potential between different groups of foreign investors, depending on their characteristics.

Data

The surveys that form the basis for this chapter were developed as part of a study carried out by the International Trade Department of the World Bank. This research aims to assist low-income countries (LICs), particularly from Sub-Saharan Africa (SSA), to take better advantage of spillovers from FDI within the context of global value chains (GVCs). Specifically, the research aims to identify the critical factors for the realization of FDI-related spillovers—including dynamic interactions between FDI and local suppliers.

The extent and nature of potential FDI-generated spillovers differ importantly by sector and FDI motive. Thus, the surveys underlying this chapter do not focus exclusively on manufacturing. Besides light manufacturing (apparel), two natural resources-based sectors that are particularly relevant for LICs in SSA were included in the dataset: mining and agribusiness. Better understanding the unique dynamics of FDI linkages and spillovers in these natural resource-intensive sectors represents an important opportunity, because they receive a significant share of FDI, particularly in developing countries. In addition, the study includes benchmark countries for these two sectors—Chile (for mining) and Vietnam (for agribusiness)—to be compared with the SSA countries.

Between March and October 2012, three different types of firms were surveyed by various consultants, namely (a) national suppliers, that is, firms with a national ownership of at least 75 percent that supply to multinationals in the country, (b) foreign investors, that is, firms that have a foreign ownership share of at least 25 percent, and (c) national producers, that is, domestic firms that are final goods producers and have a national ownership of at least 75 percent. In cases where reported data seemed unlikely, either consultants or the firms themselves were contacted again to make sure we obtained the correct numbers.

The focus of this subsection is on foreign investors, but we also compare their characteristics with domestic producers. The foreign investors' surveys cover 87 firms in Chile (5), Ghana (16), Kenya (20), Lesotho (15), Mozambique (10), Swaziland (11), and Vietnam (10). Table 4.1 shows that the majority of foreign investors are in apparel (43), followed by agribusiness (30) and mining (14). Domestic producers' surveys cover 64 firms in Chile (5), Ghana (10), Kenya (26), Mozambique (6), and Vietnam (17). The majority of these firms are in agribusiness (46), followed by apparel (13) and mining (5).

Differences between Foreign Investors and Domestic Producers

In this subsection, we assess if foreign investors are different from domestic producers in terms of their potential to generate positive spillover effects for domestic suppliers. We look at three types of indicators that influence spillover

Table 4.1 Number of Firms by Type of Firm and Sector

Type	Sector	Number of firms	Share of total (%)
Foreign investor	Agribusiness	30	34.5
Foreign investor	Apparel	43	49.4
Foreign investor	Mining	14	16.1
Foreign investor	All sectors	87	100.0
Domestic producer	Agribusiness	46	71.9
Domestic producer	Apparel	13	20.3
Domestic producer	Mining	5	7.8
Domestic producer	All sectors	64	100.0

potential, namely the firms' overall performance, their linkages with the local economy, and supplier assistance.

Performance Indicators

Table 4.2 (column 1) shows the mean differences, controlling for country-sector fixed effects. Column (2) additionally controls for employment (in natural logarithms), since firm size may also explain some of the differences between multinationals and domestic producers. All variables refer to fiscal year (FY) 2012. The results indicate that multinationals sell significantly more than domestic suppliers (ln*sales*), although the effect becomes smaller when controlling for firm size. Foreign firms are also more productive (ln*labprod*), and this effect is slightly larger when we additionally control for firm size. They also have a smaller technology gap (*tech*) to the leading domestic competitor (that is, domestic producers generally lag further behind the domestic leader in the sector), which could be the result of being more productive.

The positive coefficient sign on the share of workers with tertiary education (*emp_ter*) and the negative coefficient sign on the share of workers with

Table 4.2 Performance Indicators, Foreign Investors versus Domestic Producers (Mean Difference)

Variable	Definition	Difference (1)	Additional controls for lnemp (2)
ln*sales*	Firm's sales (US$) in natural logarithms	2.5893*** (0.000)	2.1162*** (0.000)
ln*age*	Number of years since firm has started operations in natural logarithms	−0.1429 (0.389)	−0.2192 (0.233)
ln*emp*	Firm's number of employees in natural logarithms	0.3410 (0.270)	n.a. n.a.
ln*labprod*	Firm's sales per number of employees (US$) in natural logarithms	1.9528*** (0.000)	2.1162*** (0.000)
tech	Technology gap between firm and its leading domestic competitor in the same sector, where 1 means "not existent" and 4 means "large"	−0.4982*** (0.003)	−0.6094*** (0.000)
emp_ter	Percentage of workers with tertiary education in the firm's total workforce	6.5680 (0.262)	8.9122 (0.106)
emp_sec	Percentage of workers with secondary education in the firm's total workforce	−6.7298 (0.315)	−7.8271 (0.225)
export	Dummy taking the value of 1 if a firm exports, and 0 otherwise	0.6418** (0.025)	0.5233* (0.083)
expsh_dir	Percentage of direct exports of firm's total sales	35.7146*** (0.000)	33.3476*** (0.000)
expsh_ind	Percentage of indirect exports of firm's total sales	−1.6483 (0.681)	−4.8535 (0.206)

Source: Calculations using World Bank data.
Note: Variables refer to FY 2012. All regressions control for country-sector fixed effects. Standard errors are robust to heteroscedasticity.
*$p < 0.1$, **$p < 0.05$, ***$p < 0.01$ (p values in parentheses). n.a. = not applicable.

secondary education (*emp_sec*) seem to indicate that foreign firms have a labor force that is more skilled, although the effects are not significant. Foreign firms are more likely to export (*export*). The share of direct exports is clearly higher for foreign firms (*expsh_dir*), while the share of direct exports (*expsh_ind*) shows a negative coefficient sign, but has no statistically significant impact.

In sum, we find that foreign investors tend to outperform domestic producers in terms of sales, firm size, productivity, technology gap, exporting behavior, and direct export share. This would imply a higher knowledge and productivity spillover potential compared to domestic firms.

Linkages with the Local Economy

Table 4.3 compares foreign investors' and domestic producers' linkages with the local economy. Linkages are measured in terms of the share of domestic inputs and workers as well as a firm's percentage of sales going to the domestic market. All are expected to increase the potential of positive spillovers for local suppliers (see chapter 2). We also examine differences between types of inputs and workers. We follow the specification of the previous section. All variables refer to FY 2012.

Foreign investors source a lower share of their total inputs from domestic suppliers (*inp_dom*) compared to domestic producers. We also evaluate if foreign investors and domestic producers differ in terms of their sourcing patterns. Foreign investors source a significantly lower share of raw materials (*inp_dom_mat*) and equipment and machinery (*inp_dom_equip*) as a percentage of their total domestic inputs compared to domestic producers. On the other hand, their share of technical services (*inp_dom_tech*) as well as transport, security, cleaning, catering, and other services (*inp_dom_oth*) is significantly larger in comparison with domestic producers.

We now focus on the firms' use of local workers. Foreign firms clearly employ a lower share of domestic workers (*emp_dom*) than domestic producers. The differences are slightly larger when we control for firm size (column 2). These differences are no longer statistically significant if we differentiate between types of workers by educational level. As could be expected, foreign investors significantly make less use of domestic managers (*man_dom*) compared to domestic producers. While the coefficient signs are consistently negative for supervisors (*super_dom*) and technical positions (*tech_dom*), they narrowly miss the threshold of statistical significance.

Finally, we also look at forward linkages, measured as a firm's percentage of sales going to the domestic market (*market*). The results show unambiguously that foreign investors sell a lower percentage to the local market than domestic producers.

In sum, foreign investors are characterized by fewer linkages with the local economy, as they make less use of domestic workers and inputs and also sell a lower share of their output to the domestic market. However, the findings also

Table 4.3 Linkages, Foreign Investors versus Domestic Producers (Mean Difference)

Variable	Definition	Difference (1)	Additional controls for lnemp (2)
Inputs			
inp_dom	Percentage of inputs sourced from domestic suppliers in the firm's total inputs	−16.0734*** (0.008)	−12.4843** (0.043)
inp_dom_mat	Percentage of raw materials from domestic firms of firm's total input purchases from domestic firms	−16.1221*** (0.002)	−12.4158** (0.029)
inp_dom_comp	Percentage of parts and components from domestic firms of firm's total input purchases from domestic firms	−0.1020 (0.938)	−0.3504 (0.807)
inp_dom_pack	Percentage of packaging from domestic firms of firm's total input purchases from domestic firms	3.7895 (0.331)	5.6411 (0.201)
inp_dom_equip	Percentage of equipment and machinery from domestic firms of firm's total input purchases from domestic firms	−5.0125** (0.025)	−5.0252** (0.041)
inp_dom_bus	Percentage of business services from domestic firms of firm's total input purchases from domestic firms	0.7942 (0.693)	−0.1636 (0.940)
inp_dom_tech	Percentage of technical services from domestic firms of firm's total input purchases from domestic firms	3.7713** (0.018)	3.7013** (0.031)
inp_dom_oth	Percentage of transport, security, cleaning, catering, and other services from domestic firms of firm's total input purchases from domestic firms	13.9780*** (0.000)	9.5439*** (0.001)
Labor			
emp_dom	Percentage of domestic workers in the firm's total workforce	−4.0758*** (0.002)	−4.4249*** (0.002)
emp_ter_dom	Percentage of domestic workers with tertiary education in the firm's total workforce	2.8700 (0.613)	4.1928 (0.445)
emp_sec_dom	Percentage of domestic workers with secondary education in the firm's total workforce	−7.6005 (0.261)	−8.0573 (0.225)
emp_oth_dom	Percentage of other domestic workers in the firm's total workforce	−0.1145 (0.986)	−0.7786 (0.906)
man_dom	Percentage of domestic managers of firm's total managers	−15.5842*** (0.000)	−16.4872*** (0.000)
super_dom	Percentage of domestic supervisors of firm's total supervisors	−6.6335 (0.181)	−8.5360 (0.100)
tech_dom	Percentage of technical positions of firm's total technical positions	−5.9357 (0.159)	−5.8431 (0.185)
Output			
market	Percentage of sales to domestic market of firm's total sales	−34.0663*** (0.000)	−28.4941*** (0.001)

Source: Calculations using World Bank data.
Note: Variables refer to FY 2012. All regressions control for country-sector fixed effects. Standard errors are robust to heteroscedasticity.
*$p < 0.1$, **$p < 0.05$, ***$p < 0.01$ (p values in parentheses).

show that certain service inputs, namely technical services and transport, security, cleaning, catering, and other services, show a higher potential for linkages.

Supplier Assistance

Finally, we also assess if there are differences between foreign investors and domestic producers in terms of their supplier assistance, as assistance increases the FDI spillover potential (as discussed in chapter 2). For each indicator we measure the probability of assisting suppliers, which takes the value of 1 if a firm offers assistance, and 0 otherwise. The data don't allow us to identify when and how often supplier assistance took place.

The negative coefficient signs in table 4.4 suggest that foreign investors seem to offer less assistance to local suppliers than domestic producers, although the effects are only significant for five types of assistance, namely (a) help with organization of production lines (*assist_orga*); (b) help with quality assurance (*assist_qual*); (c) help with the supplier's business strategy (*assist_strat*); (d) help with finding export opportunities (*assist_exp*), which is only significant if we control for firm size (column 2); and (v) help with implementing health, safety, environmental, and/or social conditions (*assist_hse*).

In sum, foreign investors outperform domestic producers in terms of sales, firm size, productivity, exporting behavior, and direct export share. While this would imply a higher knowledge and productivity spillover potential compared to domestic firms, foreign investors have fewer linkages with the local economy in terms of using domestic inputs and workers. There is also some evidence that foreign firms offer less assistance to local suppliers. Fewer linkages and less supplier assistance both can limit the positive impact from FDI.

Premia by Foreign Investor Characteristics

The analysis in the previous section treated foreign firms as homogenous. The literature survey in chapter 2, however, showed that certain types of FDI seem to be more beneficial than others since actual FDI spillovers also depend on foreign firm characteristics. In this section, we therefore split the foreign investors into several groups to investigate if firms with certain characteristics have a larger FDI spillover potential than others.

We estimate the following equation:

$$potential_{isc} = \alpha_0 + FC_{isc} + D_{cs} + \varepsilon_{isc} \qquad (4.1)$$

where subscript i stands for firm, s for the firm's sector, and c for country. α_0 designates the constant, D_{cs} country-sector fixed effects, and ε_{isc} the idiosyncratic error term. FC is a vector representing several foreign firm characteristics that take the value of 1 if a foreign investor fulfills a certain characteristic, and 0 otherwise. The variable *potential* is our measure of FDI spillover potential. Building on the theoretical discussion in chapter 2, we include the foreign investor characteristics shown in table 4.5.

Table 4.4 Supplier Assistance, Foreign Investors versus Domestic Producers (Mean Difference)

Variable	Definition	Difference (1)	Additional controls for lnemp (2)
assist	Dummy taking the value 1 if firm offered assistance to domestic suppliers, and 0 otherwise	−0.1725 (0.636)	−0.2994 (0.437)
assist_pay	Advance payment	−0.4019 (0.203)	−0.2117 (0.523)
assist_impr	Provision of financing for improvements	−0.3675 (0.155)	−0.4821 (0.081)
assist_funds	Support to get funds from other sources	−0.0831 (0.747)	−0.1474 (0.587)
assist_plan	Financial planning	−0.1670 (0.522)	−0.1160 (0.669)
assist_inp	Provision of inputs	−0.1683 (0.509)	−0.1846 (0.496)
assist_sourc	Support for sourcing raw materials	−0.2125 (0.405)	−0.1645 (0.544)
assist_train	Training of workers	0.0801 (0.760)	0.0111 (0.968)
assist_equip	Lending/leasing of machines or equipment	−0.0590 (0.827)	0.0247 (0.931)
assist_tech	Product or process technologies	−0.1584 (0.546)	−0.3123 (0.302)
assist_maint	Repair/maintenance of machines	−0.1376 (0.620)	−0.1472 (0.619)
assist_license	Licensing of patented technology	−0.0022 (0.994)	0.0006 (0.999)
assist_orga	Help with organization of production lines	−0.5224** (0.046)	−0.6778** (0.024)
assist_qual	Help with quality assurance	−0.5166* (0.060)	−0.5547* (0.057)
assist_invent	Help with inventory control	0.0303 (0.907)	0.0262 (0.925)
assist_audit	Help with audits	−0.1651 (0.536)	−0.1779 (0.538)
assist_strat	Help with business strategy	−0.6606** (0.012)	−0.7690*** (0.007)
assist_exp	Help with finding export opportunities	−0.4629 (0.101)	−0.5017* (0.089)
assist_hse	Help with implementing health, safety, environmental, and/or social conditions	−0.6467** (0.017)	−0.6589** (0.024)

Source: Calculations using World Bank data.
Note: All regressions control for country-sector fixed effects. Standard errors are robust to heteroscedasticity.
*$p < 0.1$, **$p < 0.05$, ***$p < 0.01$ (p values in parentheses).

Table 4.5 Foreign Investor Characteristics, Definition

Variable	Definition
own	A firm's percentage of foreign ownership
age_fdi	Number of years since a multinational has started its operations in the host country
tech	A foreign firm's technology gap with its leading domestic competitor in the same sector, where 1 means "not existent" and 4 means "large"
origin_SSA	Dummy taking the value of 1 if the largest foreign investor's region of origin is Sub-Saharan Africa (SSA), and 0 otherwise
origin_Asia	Dummy taking the value of 1 if the largest foreign investor's region of origin is Asia (including South Asia), and 0 otherwise
motive_market	Importance of access to (local and regional) markets, where 1 means "not important" and 4 means "very important"
motive_cost	Importance of access to reduced labor and nonlabor-related costs, where 1 means "not important" and 4 means "very important"
motive_res	Importance of access to raw materials and specific inputs, where 1 means "not important" and 4 means "very important"
motive_asset	Importance of access to skills and technology, where 1 means "not important" and 4 means "very important"

We apply four FDI spillover potential measures related to a foreign firm's linkages with and assistance to domestic suppliers, as these are the categories where foreign firms lag behind domestic producers: (a) the percentage of purchased goods and services sourced from domestic suppliers (*inp_dom*), (b) the percentage of domestic workers in the firm's total workforce (*emp_dom*), (c) the percentage of sales to the domestic market (*market*), and (d) the likelihood of supplier assistance (*assist*). While foreign investor characteristics refer to FY 2012, we don't know when supplier assistance took place. However, it is relatively safe to assume that major foreign characteristics remained constant over time.

Table 4.6 shows the descriptive statistics. Each line represents a foreign investor characteristic, FC, using different thresholds, while columns 1 to 4 refer to our four measures of FDI spillover potential. Each panel in a column is estimated as a separate regression.

The share of foreign ownership (*own*) matters for the FDI spillover potential. Multinationals with a foreign ownership share of at least 50 and less than 100 percent source more inputs locally compared to other firms, and this effect is even slightly higher for firms with full foreign ownership (column 1). However, we don't find any effects on alternative measures of FDI spillover potential.

A multinational's presence in the host country (*age_fdi*) is negatively associated with the share of domestically sourced inputs if the firm has been in the country for at least 20 years (column 1), but positively related with the percentage of domestic workers (column 2). A presence in the host country of at least 10 but less than 20 years is also positively related with the probability to offer supplier assistance (column 4).

If a foreign firm has a moderate technology gap (*tech*) to the leading domestic competitor in the same sector, it is more likely to offer supplier assistance (column 4).

Table 4.6 Premia by Foreign Investor Characteristics

Variable	Thresholds Foreign investor = 1 if ..., and 0 otherwise	Measure of FDI spillover potential			
		(1) inp_dom	(2) emp_dom	(3) market	(4) assist
own	50 >= own < 100%	19.3783*	0.8246	18.4457	0.7381
		(0.053)	(0.751)	(0.533)	(0.433)
	own = 100%	20.1105***	0.5891	15.6657	1.0395
		(0.006)	(0.769)	(0.575)	(0.185)
age_fdi	5 >= age_fdi < 10	−4.1679	1.1154	−5.3242	−0.4357
		(0.518)	(0.730)	(0.707)	(0.638)
	10 >= age_fdi < 20	6.3996	1.9615	−6.8739	1.5076*
		(0.176)	(0.403)	(0.487)	(0.080)
	age_fdi >= 20	−13.8976*	6.9023**	−0.8358	0.9591
		(0.055)	(0.040)	(0.965)	(0.210)
tech	tech = 2	0.6802	0.7089	20.1645	6.1271***
		(0.945)	(0.784)	(0.133)	(0.000)
	tech = 3	−1.2057	0.8178	9.5329	—
		(0.924)	(0.705)	(0.487)	—
origin	origin = SSA	2.6070	−1.2141	31.4395***	4.5044***
		(0.739)	(0.800)	(0.000)	(0.000)
	origin = Asia	−1.1053	−7.1175	30.3003***	−1.5248*
		(0.890)	(0.171)	(0.001)	(0.072)
motive_market	motive_market = 2	0.0312	−4.1798*	16.9894	—
		(0.998)	(0.075)	(0.290)	—
	motive_market >= 3	−0.4772	−2.2504	26.7538***	1.1809**
		(0.926)	(0.252)	(0.000)	(0.040)
motive_cost	motive_cost = 2	2.3507	−12.0948**	3.0408	−1.6694*
		(0.770)	(0.050)	(0.786)	(0.051)
	motive_cost >= 3	−0.9970	−3.6712	8.7955	−0.0534
		(0.877)	(0.109)	(0.440)	(0.940)
motive_res	motive_res = 2	−10.0951	−4.0206	3.0942	−5.3253***
		(0.223)	(0.292)	(0.810)	(0.000)
	motive_res >= 3	10.3145	−2.0761	−33.1588**	−10.5863***
		(0.274)	(0.509)	(0.023)	(0.000)
motive_asset	motive_asset = 2	3.7012	2.9197	4.6732	—
		(0.682)	(0.369)	(0.688)	—
	motive_asset >= 3	−5.4219	2.6393	2.5715	−0.6596
		(0.669)	(0.485)	(0.814)	(0.458)

Source: Calculations using World Bank data.
Note: All variables except for *assist* refer to FY 2012. For a description of the variables, see table 4.5. Each panel in a column is estimated as a separate regression. All regressions control for country-sector fixed effects. Standard errors are robust to heteroscedasticity. No observations are given for *tech* = 4. Cells with dashes (—) indicate variables that were dropped from the regressions.
*$p < 0.1$, **$p < 0.05$, ***$p < 0.01$ (p values in parentheses). FDI = foreign direct investment.

The region of origin (*origin*) also matters for the FDI spillover potential. Interestingly, foreign firms with the largest investor from SSA are more likely to assist domestic suppliers compared to other firms (column 4). In addition, they sell a higher share of their output to the local market (column 3). Firms with their largest foreign investor from Asia (including South Asia) also sell a significantly larger share of output to the local market, but offer significantly less assistance to their domestic suppliers (columns 3 and 4).

In a next step, we evaluate whether the FDI motive influences the extent of FDI linkages. As could be expected, market-seeking FDI (*motive_market*) is positively correlated with the share of sales to the host country (column 3). It is also positive correlated with the probability of supplier assistance (column 4). However, firms where market-seeking FDI is moderate make significantly less use of local workers (column 2).

Cost-seeking FDI (*motive_cost*) is negatively correlated with the share of local workers (column 2) as well as the probability of offering supplier assistance (column 4) if this motive has a moderate importance for multinationals.

Resource-seeking FDI (*motive_res*) clearly shows a negative correlation with the share of sales going to the host country if this motive is important (column 3). Moreover, it is also negatively associated with supplier assistance, regardless of the importance of this motive (column 4).

Which Absorptive Capacities Facilitate FDI Linkages?

This section focuses on the role of domestic supplier characteristics for FDI linkages. In the first subsection, we present the data, while the second subsection introduces the empirical model where we relate absorptive capacities with FDI linkages. The third subsection examines if there are differences in the extent of FDI linkages between different groups of suppliers, depending on their absorptive capacities. The fourth subsection describes the regression results.

Data
The focus of the following two sections is on national suppliers (see the earlier subsection "Data" for a description of our dataset). The national suppliers' surveys cover 148 firms in Chile (18), Ghana (26), Kenya (29), Mozambique (36), and Vietnam (39). More than half of the suppliers (88) supply to multinationals in agribusiness, followed by mining (48) and apparel (12). These suppliers produce a variety of inputs across the value chain, as shown in table 4.7, ranging from chemicals, to equipment, to food and food processing, to business, technical, and other services, among others.

Empirical Model
We define the following equation:

$$linkage_{isc} = \alpha_0 + AC_{isc} + D_{cs} + \varepsilon_{isc} \qquad (4.2)$$

Table 4.7 Distribution of Suppliers by Sector

Sector	Number of firms	Share of total (%)
Apparel accessories	4	2.7
Chemicals	22	14.9
Equipment	22	14.9
Food and food processing	24	16.2
Inputs to mining	8	5.4
Packaging	10	6.8
Seeds	11	7.4
Business services	17	11.5
Technical services	20	13.5
Other services	10	6.8
All sectors	148	100.0

AC is a vector denoting supplier-specific absorptive capacities that facilitate FDI linkages, and *linkage* is our measure of FDI linkages. Building on the theoretical discussion in chapter 2, we include the following absorptive capacities, as defined in table 4.8.

$$outp_{isc} = \alpha_0 + gap_{isc} + soph_{isc} + emp\_ter_{isc} + emp\_sec_{isc}$$
$$+ \ln exper_{isc} + man\_educ_{isc} + man\_exper_{isc}$$
$$+ \ln emp_{isc} + export_{isc} + \ln dist_{isc} + D_{cs} + \varepsilon_{isc} \quad (4.3)$$

Due to lack of data on research and development (R&D) activity, we use *soph* as a proxy. The variables *emp_ter* and *emp_sec* serve as our direct measures of worker skills. The variable *exper* measures a supplier's experience and thus serves as an indirect measure of skills. We also include characteristics related to the skills and experience of the general manager, namely *man_educ* and *man_exper*. The variable *emp* captures firm size, *export* a firm's export activity, and *dist* firm location. We also include a measure of technology gap (rather than firm-level productivity per se), *gap*, as has been outlined in the literature.

Since the supplier characteristics refer to the survey year (2012), we are constrained to use a linkage measure of the same year. We use the percentage of a supplier's output to foreign customers (*outp*). While *outp* does not capture direct productivity gains or other FDI spillovers, a higher share of output to foreign customers makes positive spillovers, for instance via assistance or requirements from the multinational, more likely.

Supplier Premia by Absorptive Capacity

In this section, we split suppliers into several groups to investigate if suppliers with certain characteristics benefit from larger FDI linkages than others. Modifying the specification of equation (4.2), we assign a dummy taking the

Table 4.8 Definition of Supplier Characteristics

Variable	Definition
gap	Technology gap to the leading domestic competitor's technology in the firm's sector, ranging from 1 to 4, where 1 means "no difference" and 4 means "large difference"
soph	Degree of sophistication of the firm's production process, ranging from 1 to 4, where 1 means "standardized" and 4 means "highly sophisticated"
emp_ter	Percentage of workers with tertiary education in the firm's workforce
emp_sec	Percentage of workers with secondary education in the firm's workforce
exper	Number of years since firm has started operations in country
man_educ	Highest level of education of the general manager, ranging from 1 to 3, where 1 means "primary education (without vocational education)"; 2 means "secondary education (vocational education and training)"; and 3 means "tertiary education (college or university degree)"
man_exper	Dummy taking the value of 1 if the general manager has previous work experience in a foreign firm in the country or abroad, and 0 otherwise
emp	Total employment
export	Dummy taking the value of 1 if a firm exports, and 0 otherwise
dist	Geographical distance of firm to foreign client in kilometers

value of 1 for suppliers with a certain absorptive capacity, AC, and 0 for all other suppliers in the sample, and estimate the impact on the percentage of a supplier's output to foreign customers (*outp*).

Table 4.9 shows the descriptive statistics. Each line represents a supplier's absorptive capacity, AC, applying different thresholds. Each panel is estimated as a separate regression. A highly sophisticated production process (*soph*) has a significantly positive impact on suppliers' output to foreign firms. Moreover, FDI linkages tend to increase with a more sophisticated production process, as can be seen by the growing coefficient signs on *soph* and the decreasing p values.

Firms with a share of workers with secondary education (*emp_sec*) of at least 20 and below 50 percent supply a significantly higher share to foreign investors than other firms. This effect becomes slightly smaller for suppliers employing at least 50 but less than 80 percent of workers with secondary education. However, the effect is no longer significant for suppliers with a share of workers with secondary education of at least 80 percent. The results imply that multinationals in our sample source inputs from domestic suppliers that are somewhat but not too skill intensive.

Firm size also has an influence on FDI linkages. Suppliers with at least 50 but less than 250 employees have a significantly lower output share than other suppliers. The effect is also negative for alternative threshold levels, but misses the levels of statistical significance narrowly.

Finally, geographical location also matters. FDI linkages are significantly lower for suppliers that are located more than 500 kilometers from their foreign clients (*dist*), but the negative effect levels off for suppliers that are located closer to their foreign client. Given the existence of premia for several supplier groups, we assess the impact of supplier characteristics on the extent of FDI linkages in the next section.

Table 4.9 Supplier Premia by Absorptive Capacity

Variable	Thresholds Supplier = 1 if ..., and 0 otherwise	Measure of FDI linkage: outp	
		Difference	p value
gap	gap = 2	−7.2833	(0.448)
	gap >= 3	−2.8160	(0.713)
soph	soph = 2	0.7105	(0.941)
	soph = 3	5.7639	(0.516)
	soph = 4	23.1604*	(0.072)
emp_ter	20 >= emp_ter < 50	12.6626	(0.112)
	50 >= emp_ter < 80	−5.5474	(0.541)
	emp_ter >= 80	−8.9682	(0.526)
emp_sec	20 >= emp_sec < 50	18.2152**	(0.042)
	50 >= emp_sec < 80	15.5753*	(0.095)
	emp_sec >= 80	8.4187	(0.484)
exper	3 >= exper < 10	20.9871	(0.139)
	10 >= exper < 20	14.4016	(0.296)
	20 >= exper < 30	6.4514	(0.647)
	exper >= 30	27.5507*	(0.080)
man_educ	man_educ = 2	3.3842	(0.841)
	man_educ = 3	−10.1846	(0.493)
man_exper	man_exper = 1	7.3526	(0.314)
emp	10 >= emp < 50	−18.1670	(0.157)
	50 >= emp < 250	−24.1310*	(0.072)
	emp >= 250	−23.7696	(0.118)
export	export = 1	9.8261	(0.121)
dist	20 >= dist < 100	−19.9154*	(0.056)
	100 >= dist < 500	−18.0726*	(0.057)
	dist >= 500	−26.1891***	(0.005)

Source: Calculations using World Bank data.
Note: All variables refer to FY 2012. For a description of the variables, see table 4.8. Each panel is estimated as a separate regression. All regressions control for country-sector fixed effects. Standard errors are robust to heteroscedasticity. FDI = foreign direct investment.
*$p < 0.1$, **$p < 0.05$, ***$p < 0.01$ (p values in parentheses).

Regression Results

Overall Results

Table 4.10 reports the regression results based on the specification of equation (4.3). Given the differences between supplier sectors and countries, all regressions control for country-sector fixed effects. Standard errors are robust to heteroscedasticity. A more sophisticated production process (*soph*) has a significantly positive impact on suppliers' output to foreign firms, supporting the positive role of research and development (R&D) for local firms in the literature. Firm location also matters for FDI linkages. A larger distance to the foreign firm (ln*dist*) reduces the supplier's output share going to foreign clients, supporting the findings of Barrios, Luisito, and Strobl (2006), who find evidence that foreign firms collocating in the same sector and region significantly increase productivity and employment. A larger size (ln*emp*) seems to be negatively associated with FDI

Table 4.10 The Effect of Suppliers' Absorptive Capacity on Output Share to Foreign Firms

Dependent variable: $outp_{isc}$

	(1)	(2)	(3)	(4)	(5)	(6)	(7)	(8)	(9)
gap_{isc}	−1.6276								−1.0695
	(0.609)								(0.800)
$soph_{isc}$	5.9014*								6.4544
	(0.094)								(0.120)
emp_ter_{isc}		−0.1314							−0.2208
		(0.317)							(0.234)
emp_sec_{isc}			0.1005						0.0037
			(0.396)						(0.980)
$lnexper_{isc}$				1.4755					4.7960
				(0.744)					(0.450)
man_educ_{isc}					−10.0299				−6.3287
					(0.142)				(0.412)
man_exper_{isc}					6.0535				9.7105
					(0.419)				(0.283)
$lnemp_{isc}$						−3.4974			−6.7818*
						(0.106)			(0.051)
$export_{isc}$							9.8261		10.2026
							(0.121)		(0.296)
$lndist_{isc}$								−4.0871**	−2.9573*
								(0.014)	(0.069)
$constant_{isc}$	48.7270**	63.7402***	54.2755***	54.3062**	83.3056***	70.1656***	56.2935***	69.4351***	80.0081***
	(0.013)	(0.001)	(0.002)	(0.013)	(0.000)	(0.000)	(0.001)	(0.000)	(0.003)
Country-sector FE	Yes	Yes	Yes	Yes	Yes	Yes	Yes	Yes	Yes
R^2	0.31	0.32	0.31	0.29	0.33	0.33	0.30	0.34	0.48
Observations	109	107	107	109	112	107	110	105	93

Source: Calculations using World Bank data.

Note: All variables refer to FY 2012. For a description of the variables, see table 4.8. All regressions control for country-sector fixed effects. Standard errors are robust to heteroscedasticity. FE = fixed effects; OLS = ordinary least squares.

*$p < 0.1$, **$p < 0.05$, ***$p < 0.01$ (p values in parentheses).

linkages, while exporting (*export*) seems to have a positive impact, although both narrowly miss the 10 percent threshold of statistical significance. Including all absorptive capacities simultaneously (column 9) confirms the findings only for firm size (ln*emp*) and distance to the foreign firm (ln*dist*).

Results for Established Suppliers

It is likely that firms having a longer supplier experience have different absorptive capacities compared to firms that just started supplying to a foreign client, especially as structural changes (such as changes in the supplier's capacity, sophistication of production processes, or skill levels) may happen early on during their relationship. We therefore rerun the regressions for supplier firms that have a supplier relationship of at least three years (see table 4.11).

We can confirm the positive impact of a more sophisticated production process (*soph*) and the negative impact of a larger distance to the foreign firm (ln*dist*). However, we also find a significantly negative impact of the share of workers with tertiary education (*emp_ter*) on the supplier's share of output going to foreign firms. A higher educational level of the general manager (*man_educ*) also reduces FDI linkages. While our focus here is on the suppliers' output share to foreign firms and not on FDI spillovers, our findings can be related to those by Sinani and Meyer (2004), who find that a larger share of human capital leads to negative FDI spillovers (see chapter 2), although the underlying mechanisms may be different. It may be possible that suppliers with highly educated managers supply a larger share of inputs to firms abroad, for instance, because they may have fewer language barriers. The combined results (column 9) indicate that the negative effect of human capital is no longer significant when we control for a firm's exporting activity. Only distance to the foreign firm (ln*dist*) shows a significant effect.

Which Factors within Transmission Channels Support FDI Spillovers?

Supplier Premia by Factors within Transmission Channel

In this section, we evaluate whether suppliers that benefited from any demand or assistance effects are characterized by higher FDI linkages and spillovers than suppliers that don't. Table 4.12 shows the supplier premia by transmission channel. Focusing on FDI linkages first, firms that received assistance from the foreign customer to make improvements (*assist*) supply a significantly higher share of their output to foreign clients than firms that don't (*outp* column).

Besides FDI linkages, we also include *exp_start* as our FDI spillover measure, which is a dummy taking the value of 1 if the firm started exporting as a consequence of supplying to a foreign customer, and 0 otherwise. The results confirm that several transmission channels matter for backward FDI spillovers. Suppliers receiving technical audits before or after signing the contract (*audit*), suppliers receiving assistance from their foreign clients (*assist*), suppliers with joint product development with their customers (*dev*), and suppliers licensing technology from their foreign client (*license*) are more likely to export as a result of their supplier relationship (*exp_start* column). In sum, we find evidence for

Table 4.11 The Effect of Suppliers' Absorptive Capacity with Supplier Relationship of at Least Three Years on Output Share to Foreign Firms

Dependent variable: $outp_{isc}$

	(1)	(2)	(3)	(4)	(5)	(6)	(7)	(8)	(9)
gap_{isc}	−3.1899								−3.4042
	(0.341)								(0.412)
$soph_{isc}$	6.9340*								6.8075
	(0.055)								(0.102)
emp_ter_{isc}		−0.2337*							−0.2822
		(0.067)							(0.105)
emp_sec_{isc}			0.1797						−0.0169
			(0.179)						(0.911)
$lnexper_{isc}$				−2.2745					−0.0370
				(0.709)					(0.996)
man_educ_{isc}					−14.2539**				−13.6016
					(0.048)				(0.176)
man_exper_{isc}					5.5674				10.0019
					(0.469)				(0.265)
$lnemp_{isc}$						−2.3064			−4.5781
						(0.302)			(0.200)
$export_{isc}$							10.8120		7.7413
							(0.114)		(0.414)
$lndist_{isc}$								−3.7772**	−2.5183*
								(0.025)	(0.097)
$constant_{isc}$	49.5283**	67.6262***	50.7481***	65.6000**	95.1645***	66.2781***	56.0470***	68.6250***	113.8197***
	(0.015)	(0.001)	(0.008)	(0.014)	(0.000)	(0.000)	(0.001)	(0.000)	(0.001)
Country-sector FE	Yes	Yes	Yes	Yes	Yes	Yes	Yes	Yes	Yes
R^2	0.33	0.36	0.34	0.30	0.35	0.33	0.32	0.36	0.54
Observations	102	100	100	102	105	100	103	99	87

Source: Calculations using World Bank data.

Note: All variables refer to FY 2012. For a description of the variables, see table 4.8. All regressions control for country-sector fixed effects. Standard errors are robust to heteroscedasticity. FE = fixed effects; OLS = ordinary least squares.

*$p < 0.1$, **$p < 0.05$, ***$p < 0.01$ (p values in parentheses).

Table 4.12 Supplier Premia by Factors within Transmission Channels

Variable	Definitions	Measure: outp	Measure: exp_start
audit	Dummy taking the value of 1 if supplier received technical audits before or after signing a contract with the foreign customer, and 0 otherwise.	−0.6666 (0.909)	0.8551** (0.049)
impr	Dummy taking the value of 1 if the foreign customer required the supplier to make improvements before or after signing the contract, and 0 otherwise.	1.9031 (0.796)	0.3366 (0.468)
assist	Dummy taking the value of 1 if supplier received assistance from the foreign customer to meet any requirements before or after signing the contract, and 0 otherwise.	16.5684** (0.013)	1.3256*** (0.008)
dev	Dummy taking the value of 1 if supplier developed product jointly with the foreign customer, and 0 otherwise.	10.7522 (0.129)	1.2506*** (0.006)
license	Dummy taking the value of 1 if supplier licensed technology from the foreign customer, and 0 otherwise.	5.1151 (0.498)	1.2387** (0.014)

Source: Calculations using World Bank data.
Note: All regressions control for country-sector fixed effects. Standard errors are robust to heteroscedasticity.
$*p < 0.1$, $**p < 0.05$, $***p < 0.01$ (p values in parentheses).

the existence of positive assistance effects in the narrower and wider sense (including assistance, technical audits, joint product development, and technology licensing) in GVCs, while demand effects (measured as requirements to improve) do not have any impact. We have to emphasize that it may be difficult to disentangle demand and assistance effects, as assistance effects may actually derive from demands.

Empirical Model

In this second exercise, we focus on the role of transmission channels for FDI spillovers:

$$spillover_{isc} = \alpha_0 + TC_{isc} + D_{cs} + \varepsilon_{isc} \qquad (4.4)$$

The term *TC* is a vector relating to various factors within transmission channels through which multinationals influence national suppliers and thus make FDI spillovers more likely, and *spillover* is our measure of FDI spillover.

We specify the following transmission channels, as defined in the previous subsection:

$$spillover_{isc} = \alpha_0 + audit_{isc} + impr_{isc} + assist_{isc} + dev_{isc} + license_{isc} + D_{cs} + \varepsilon_{isc} \qquad (4.5)$$

impr captures demand effects in GVCs, while *audit*, *assist*, *dev*, and *license* represent assistance effects. We use *exp_start* as our spillover measure (see the previous subsection for a definition).

Regression Results

Overall Results

Table 4.13 follows the specification of equation (4.5) and uses exporting as a consequence of supplying to a foreign customer (*exp_start*) as the spillover measure. Technical audits (*audit*), assistance by foreign customers (*assist*), joint product development (*dev*), and licensed technology from the foreign customer (*license*) all significantly influence a supplier's likelihood of starting to export as a result of supplying to a foreign customer. In the combined sample (column 6), we can confirm the significantly positive effects of technical audits (*audit*) and assistance by foreign customers (*assist*). Again, requirements to improve (*impr*) do not have any impact, supporting our previous finding of no demand effects.

Results by Types of Requirements

The nonexistence of demand effect, that is, spillovers from a customer's requirements to improve (*impr*), raises the question whether only specific types of requirements to improve may be relevant to FDI spillovers. Using the specification of equation (4.5), we substitute 13 subindicators for *impr*,[3] which take the value of 1 if the foreign customer required the supplier to make improvements before or after signing the contract, and 0 otherwise. Of the 13 subindicators of *impr*, none shows a significant impact (see Winkler 2013 for detailed results). In sum, the regression results give evidence of strong assistance effects in GVCs,

Table 4.13 The Effect of Factors within Transmission Channels on the Probability of Starting to Export, Probit

Dependent variable: exp_start_{isc}						
	(1)	(2)	(3)	(4)	(5)	(6)
$audit_{isc}$	0.8551**					0.9166*
	(0.049)					(0.071)
$impr_{isc}$		0.3366				−0.1203
		(0.468)				(0.827)
$assist_{isc}$			1.3256***			1.4075***
			(0.008)			(0.008)
dev_{isc}				1.2506***		0.8537
				(0.006)		(0.138)
$license_{isc}$					1.2387**	0.8975
					(0.014)	(0.105)
$constant_{isc}$	−6.9418***	−6.4233***	−6.0867***	−7.3373***	−6.0867***	−7.7367***
	(0.000)	(0.000)	(0.000)	(0.000)	(0.000)	(0.000)
Country-sector FE	Yes	Yes	Yes	Yes	Yes	Yes
Adjusted R^{2a}	−0.219	−0.267	−0.161	−0.172	−0.197	−0.121
Observations	55	55	55	55	55	55

Source: Calculations using World Bank data.
Note: For a description of the variables, see table 4.12. All regressions control for country-sector fixed effects. Standard errors are robust to heteroscedasticity. FE = fixed effects.
a. McFadden's adjusted pseudo R^2.
*$p < 0.1$, **$p < 0.05$, ***$p < 0.01$ (p values in parentheses).

but no evidence of demand effects. One possible explanation may be the lower level of development in our country sample, where foreign investors cannot expect local suppliers to make any improvements on their own, but are rather required to offer assistance. A second explanation may be that assistance effects result from underlying demand effects (for example, where consumer standards compel lead firms to give assistance to suppliers in order to ensure compliance).

Results by Types of Assistance

In this section, we study in more detail which types of assistance are most effective in generating positive FDI spillovers in our data sample. Table 4.4 shows the definitions of the different subindicators of *assist* available in the dataset. Again, assistance is measured as a dummy taking the value of 1 if a supplier obtains assistance from the multinational, and 0 otherwise. Table 4.14 and table 4.15 report the results using the specification of equation (4.5). The tables substitute various types of assistance for *assist* and use the likelihood to start exporting due to a supplier relationship with a foreign customer (*exp_start*) as the dependent variable.

Ten types of assistance significantly increase the likelihood to start exporting as a consequence of supplying to foreign firms, namely (a) advance payment (*assist_pay*), (b) provision of financing for improvements (*assist_impr*), (c) support for sourcing raw materials (*assist_sourc*), (d) training of workers (*assist_train*), (e) product or process technologies (*assist_tech*), (f) licensing of patented technology (*assist_license*), (g) help with the organization of production lines (*assist_orga*), (h) help with quality assurance (*assist_qual*), (i) help with finding export opportunities (*assist_exp*), and (j) help with implementing health, safety, environmental, and/or social conditions (*assist_hse*). Overall, all types of assistance show a positive coefficient sign, and many miss the threshold level of statistical significance only narrowly. In sum, we find strong evidence of assistance effects in GVCs for FDI spillovers.

Table 4.14 The Effect of Assistance on the Probability of Starting to Export due to Relationship with Foreign Firm, Part 1, Probit

Dependent variable: exp_start_{lsc}

	(1)	(2)	(3)	(4)	(5)	(6)	(7)	(8)	(9)
$audit_{lsc}$	0.9181*	0.9638*	0.9207*	0.9022*	0.8890*	1.0122*	0.9072*	0.9092*	0.9766*
	(0.070)	(0.055)	(0.071)	(0.071)	(0.077)	(0.062)	(0.073)	(0.068)	(0.051)
$impr_{lsc}$	−0.2019	0.2380	−0.0289	0.0364	−0.1210	−0.1012	−0.1148	0.0158	−0.0980
	(0.712)	(0.659)	(0.955)	(0.945)	(0.817)	(0.845)	(0.824)	(0.976)	(0.853)
dev_{lsc}	0.6726	0.4277	0.7549	0.8038	0.9490*	0.8419	0.8734*	0.7910	0.3870
	(0.221)	(0.458)	(0.185)	(0.127)	(0.061)	(0.102)	(0.084)	(0.130)	(0.509)
$license_{lsc}$	0.8968*	0.5970	0.8004	0.7349	0.6149	0.8788*	0.5805	0.6898	0.7940
	(0.097)	(0.324)	(0.159)	(0.191)	(0.277)	(0.092)	(0.305)	(0.223)	(0.187)
$assist\_pay_{lsc}$	1.1684**								
	(0.024)								
$assist\_impr_{lsc}$		1.7908**							
		(0.026)							
$assist\_funds_{lsc}$			0.8546						
			(0.286)						
$assist\_plan_{lsc}$				0.9034					
				(0.210)					
$assist\_inp_{lsc}$					0.9644				
					(0.143)				
$assist\_sourc_{lsc}$						1.1450*			
						(0.083)			
$assist\_train_{lsc}$							1.2032*		
							(0.067)		
$assist\_equip_{lsc}$								0.9497	
								(0.160)	

table continues next page

Table 4.14 The Effect of Assistance on the Probability of Starting to Export due to Relationship with Foreign Firm, Part 1, Probit *(continued)*

Dependent variable: exp_start_{isc}

	(1)	(2)	(3)	(4)	(5)	(6)	(7)	(8)	(9)
$assist\_tech_{isc}$									1.6031**
									(0.020)
$constant_{isc}$	−7.4756***	−7.7162***	−7.7334***	−7.8291***	−7.8037***	−7.8395***	−7.7525***	−7.8026***	−7.3524***
	(0.000)	(0.000)	(0.000)	(0.000)	(0.000)	(0.000)	(0.000)	(0.000)	(0.000)
Country-sector FE	Yes	Yes	Yes	Yes	Yes	Yes	Yes	Yes	Yes
Adjusted R^{2a}	−0.159	−0.163	−0.205	−0.202	−0.197	−0.179	−0.184	−0.199	−0.163
Observations	55	55	55	55	55	55	55	55	55

Source: Calculations using World Bank data.
Note: For a description of the variables, see tables 4.4 and 4.12. All regressions control for country-sector fixed effects. Standard errors are robust to heteroscedasticity. FE = fixed effects.
a. McFadden's adjusted pseudo R^2.
*$p < 0.1$, **$p < 0.05$, ***$p < 0.01$ (p values in parentheses).

Table 4.15 The Effect of Assistance on the Probability of Starting to Export due to Relationship with Foreign Firm, Part 2, Probit

Dependent variable: exp_start_{isc}

	(1)	(2)	(3)	(4)	(5)	(6)	(7)	(8)	(9)
$audit_{isc}$	0.8834*	0.9558*	0.8693*	0.7906*	0.8267*	0.8267*	0.8751*	0.7924*	1.0472*
	(0.079)	(0.061)	(0.075)	(0.099)	(0.087)	(0.087)	(0.079)	(0.099)	(0.053)
$impr_{isc}$	0.0178	0.0100	−0.1796	0.0824	−0.1168	−0.1168	−0.1327	−0.1387	−0.1363
	(0.973)	(0.985)	(0.740)	(0.875)	(0.826)	(0.826)	(0.802)	(0.794)	(0.798)
dev_{isc}	0.8849*	0.8012	0.6690	0.8825*	0.9440*	0.9440*	0.8656*	0.8684*	0.6131
	(0.081)	(0.134)	(0.236)	(0.091)	(0.065)	(0.065)	(0.089)	(0.099)	(0.258)
$license_{isc}$	0.6547	0.7734	0.7828	0.6869	0.6901	0.6901	0.7473	0.5457	0.7957
	(0.254)	(0.141)	(0.179)	(0.236)	(0.228)	(0.228)	(0.185)	(0.330)	(0.136)
$assist\_maint_{isc}$	0.6738								
	(0.260)								
$assist\_license_{isc}$		1.4250**							
		(0.016)							
$assist\_orga_{isc}$			1.0601*						
			(0.060)						
$assist\_qual_{isc}$				1.0160**					
				(0.041)					
$assist\_invent_{isc}$					0.6007				
					(0.387)				
$assist\_audit_{isc}$						0.6007			
						(0.387)			
$assist\_strat_{isc}$							0.6723		
							(0.145)		
$assist\_exp_{isc}$								1.2943**	
								(0.027)	

table continues next page

Table 4.15 The Effect of Assistance on the Probability of Starting to Export due to Relationship with Foreign Firm, Part 2, Probit *(continued)*

Dependent variable: exp_start$_{fsc}$

	(1)	(2)	(3)	(4)	(5)	(6)	(7)	(8)	(9)
assist_hse$_{fsc}$									1.4993**
									(0.014)
constant$_{fsc}$	−7.8728***	−7.8537***	−7.4454***	−7.8423***	−7.7406***	−7.7406***	−7.6948***	−7.6089***	−7.6106***
	(0.000)	(0.000)	(0.000)	(0.000)	(0.000)	(0.000)	(0.000)	(0.000)	(0.000)
Country-sector FE	Yes	Yes	Yes	Yes	Yes	Yes	Yes	Yes	Yes
Adjusted R^2[a]	−0.204	−0.150	0.183	−0.183	−0.209	−0.209	−0.199	−0.179	−0.139
Observations	55	55	55	55	55	55	55	55	55

Source: Calculations based on World Bank data.
Note: For a description of the variables, see tables 4.4 and 4.12. All regressions control for country-sector fixed effects. Standard errors are robust to heteroscedasticity. FE = fixed effects.
a. McFadden's adjusted pseudo R^2.
*$p < 0.1$, **$p < 0.05$, ***$p < 0.01$ (p values in parentheses).

Conclusions

Summary of Results

Using newly collected survey data on direct supplier-multinational linkages in Chile, Ghana, Kenya, Lesotho, Mozambique, Swaziland, and Vietnam, this chapter evaluated whether foreign investors differ from domestic producers in terms of their overall performance, linkages with the local economy, and supplier assistance, which all influence the foreign firms' potential to generate productivity spillovers. Besides apparel, the firms in our sample cover two natural-resource-intensive industries, namely agribusiness and mining. We found that foreign investors outperform domestic producers in terms of sales, firm size, productivity, exporting behavior, and direct export share. This would imply a higher knowledge and productivity spillover potential compared to domestic firms, but foreign investors have fewer linkages with the local economy in terms of using domestic inputs and workers. However, the findings also show that certain service inputs—namely technical services and transport, security, cleaning, catering, and other services—show a higher potential for linkages. There is also some evidence that foreign firms offer less assistance to local suppliers. Fewer linkages and supplier assistance both can limit the positive impact from FDI.

In a next step, we studied the relationship between foreign investor characteristics and the FDI spillover potential. In sum, we found that foreign investor characteristics matter for FDI linkages and supplier assistance, but the size and direction of the relationship depends on the measure of FDI spillover potential we used. For example, a multinational's presence in the host country is negatively associated with the share of domestically sourced inputs if the firm has been in the country for at least 20 years, but positively related with the percentage of domestic workers. Other foreign firm characteristics, on the other hand, show a less ambiguous picture. Market-seeking FDI, for example, shows a positive relationship with the share of sales to the host country as well as the probability of supplier assistance. And foreign firms with the largest investor from SSA are associated with a larger share of sales to the local market and a higher likelihood of supplier assistance. Foreign firms with the largest investor from Asia also sell a significantly larger share of output to the local market, but offer significantly less assistance to their domestic suppliers.

The second part of this chapter first examined the role of supplier firms' absorptive capacities for FDI linkages. These firms supply to multinationals in agribusiness, mining, and apparel, but produce a variety of inputs across the value chain. The results indicated that several supplier characteristics matter for FDI linkages, measured as the share of output going to multinationals, which in turn increases the FDI spillover potential. A more sophisticated production process has a significantly positive impact on FDI linkages, whereas a larger geographic distance to the foreign client shows a negative effect. The descriptive statistics also showed that firms with a share of workers with secondary education of at least 20 percent supply a significantly higher share to foreign investors than other firms. Although this effect could not be confirmed by the regression results covering the full sample, we found a significantly negative impact of the share of workers with tertiary education on FDI linkages when we focus on suppliers with a supplier relationship of at least three years. The general manager's educational level also has a negative effect. Overall, these findings suggest that a larger share of human capital leads to reduced FDI linkages in supplier firms. One possible explanation for this unexpected result could be that suppliers with highly educated managers supply a larger share of inputs to firms abroad, for instance, because they may have fewer language barriers. Finally, we also found evidence that a higher number of employees reduce the supplier's share of output to foreign firms.

In a next step, we assessed whether factors within the transmission channels between multinationals and suppliers influence FDI spillovers, focusing on assistance and demand effects. We used exporting as a consequence of supplying to a foreign customer as our spillover measure. The results confirmed that several transmission channels matter for backward FDI spillovers. Suppliers receiving technical audits before or after signing the contract, suppliers receiving assistance from their foreign clients, suppliers with joint product development with their customers, and suppliers licensing technology from their foreign client are more likely to export as a result of their supplier relationship. In sum, we find evidence for the existence of positive assistance effects (including technical audits, joint product development, and technology licensing) in GVCs, while demand effects (measured as requirements to improve) do not have any impact.

Finally, we also studied which types of assistance are most effective in generating positive FDI spillovers in our data sample. Ten types of assistance significantly increase the likelihood to start exporting as a consequence of supplying to foreign firms, namely advance payment, provision of financing for improvements, support for sourcing raw materials, training of workers, product or process technologies, licensing of patented technology, help with the organization of production lines, help with quality assurance, help with finding export opportunities, and help with implementing health, safety, environmental, and/or social conditions.

Policy Conclusions

Our findings suggest that the FDI spillover potential via GVCs depends on the extent, durability, and quality of linkages between foreign investors and the local economy. Investment promotion alone is not sufficient to benefit from FDI spillovers. It is important to embed foreign investors into the local economy to increase the amount and quality of linkages, and therefore the possibility for supplier assistance and the potential for FDI spillovers in the long term. In order to integrate foreign investors into local value chains, government agencies could identify potential domestic suppliers and encourage foreign investors to participate in supplier development and assistance. Governments could also give incentives to multinationals to collaborate with local universities, research institutes, or other firms, which would improve the local skill and innovation capacity (Potter 2002).

Policies that aim at increasing FDI linkages will be more targeted if foreign firm characteristics and the absorptive capacities of domestic suppliers are taken into account. Our results have shown, for example, that the foreign investor's origin and investment motive as well as the share of foreign ownership matter for FDI linkages and supplier assistance. In addition, policies should aim at strengthening absorptive capacities that have been shown to increase FDI linkages, including the degree of sophistication of suppliers' production processes. Policies should also target some of the obstacles to FDI linkages, such as large geographic distances between suppliers and their foreign clients. Removing barriers to natural agglomeration, for example, through investments in infrastructure, the provision of social services, or regional integration arrangements, could reduce geographic distances between suppliers and multinationals and thus increase the FDI spillover potential.

Finally, researchers should focus more strongly on understanding better the transmission channels leading to FDI spillovers. While our chapter focused on assistance and demand effects, other transmission channels in value chains include diffusion, availability, and quality effects. Besides transmission channels in value chains, research also needs to explore better the effect of changing market forces (demonstration and competition effects) and labor turnover. This will help guide policies designed to remove barriers within transmission channels, enabling the FDI spillover potential to translate into actual FDI spillovers.

Notes

1. With substantial input from Beata Javorcik.
2. A more detailed version of this chapter is published as: D. Winkler, 2013, "Potential and Actual FDI Spillovers in the Global Value Chain: The Role of Foreign Investor Characteristics, Absorptive Capacity and Transmission Channels," Policy Research Working Paper 6424, World Bank, Washington, DC.
3. These include requirements to reorganize the product lines; to invest in new equipment and/or technology; to improve product quality, quality control, productivity, timeliness of delivery, inventory management, business management, health, safety, environmental, and/or social conditions; to increase volume of production; to cut waste; to acquire ISO 9000 or 14000; and to train employees.

References

Barrios, S., B. Luisito, and E. Strobl. 2006. "Coagglomeration and Spillovers." *Regional Science and Urban Economics* 36 (4): 467–81.

Gentile-Lüdecke, S., and A. Giroud. 2012. "Knowledge Transfer from MNCs and Upgrading of Domestic Firms: The Polish Automotive Sector." *World Development* 40 (4): 796–807.

Giroud, A., B. Jindra, and P. Marek. 2012. "Heterogeneous FDI in Transition Economies: A Novel Approach to Assess the Developmental Impact of Backward Linkages." *World Development* 40 (11): 2206–20.

Javorcik, B., and M. Spatareanu. 2009. "Tough Love: Do Czech Suppliers Learn from Their Relationships with Multinationals?" *Scandinavian Journal of Economics* 111 (4): 811–33.

Jordaan, J. 2011. "Local Sourcing and Technology Spillovers to Mexican Suppliers: How Important Are FDI and Supplier Characteristics?" *Growth and Change* 42 (3): 287–319.

Potter, J. 2002. "Embedding Foreign Direct Investment." LEED Programme, Territorial Development Service, OECD, Paris.

Sinani, E., and K. Meyer. 2004. "Spillovers of Technology Transfer from FDI: The Case of Estonia." *Journal of Comparative Economics* 32 (3): 445–66.

Winkler, D. 2013. "Potential and Actual FDI Spillovers in Global Value Chains—The Role of Foreign Investor Characteristics, Absorptive Capacity and Transmission Channels." Policy Research Working Paper 6424, World Bank, Washington, DC.

PART 3

Sector Case Studies

Sector Case Study: Mining   117
Sector Case Study: Agribusiness   163
Sector Case Study: Apparel   209

CHAPTER 5

Sector Case Study: Mining

Kaiser Associates Economic Development Partners[1]

Abstract

For many low-income countries, particularly in Sub-Saharan Africa, the mining sector represents one of the most crucial sources of investment and income in their economies. This sector relies heavily on foreign investment, and foreign direct investment (FDI) inflows have expanded rapidly over the past decade. Linkages and spillovers from this FDI can play a critical role in ensuring that the countries benefit over and beyond the lifespan of a mining project and develop sustainable and competitive alternative sectors. Given the nonrenewable nature of the mining sector, this is a particularly critical concern for policy makers. This case study shows that while current linkages—in supply chains, labor markets, and wider networks—remain limited, progress is being made, and scope exists for achieving deeper local economy integration, and ultimately productivity-enhancing spillovers. Indeed, the Chilean case provides an example of how to leverage the potential of the mining sector, and the Ghanaian case shows that some African countries are making progress down that path, if still slowly. Improving the potential for spillovers requires deep efforts in building supply-side capacity (most of which are not related directly to mining). But it also requires an active, collaborative sectoral strategy, combined with a credible and efficient regulatory approach.

The Context for Mining FDI Spillovers

Overview of Mining Global Value Chains

Probably the key difference between mining and the other sectors is that mining is a nonrenewable, finite resource with mining projects typically lasting between 15 and 25 years, although in some cases they can be more long term. As a result, local actors and governments in host countries have higher expectations and are more motivated to ensure positive spillovers from foreign direct investment (FDI), which has consequences for the value chain governance.

The mining value chain includes two main stages—mine development (acquisition of a mining concession and mine construction) and mine operations (mineral extraction and beneficiation). Core mining companies are involved in the mining value chain but do not typically invest downstream in nonmining manufacturing. Their core competences are in mine planning, mineral extraction, and basic processing. The value chain also increasingly includes end-of-life mine closure, as well as rehabilitation and alternative use/recycling of metals. Figure 5.1 highlights the segments of the value chain and the main inputs used in each.

Different commodities often exhibit different value chain structures, depending on the nature/value of the mineral along the stages of the value chain. In terms of beneficiation or value addition by the mining sector, mining companies are typically involved in basic processing, but rarely process intermediate products into semimanufactured or manufactured products suitable for industry use. This can mean that opportunities for spillovers linked to downstream beneficiation are limited in developing countries.

Figure 5.1 Mining Value Chain and the Main Inputs across the Chain

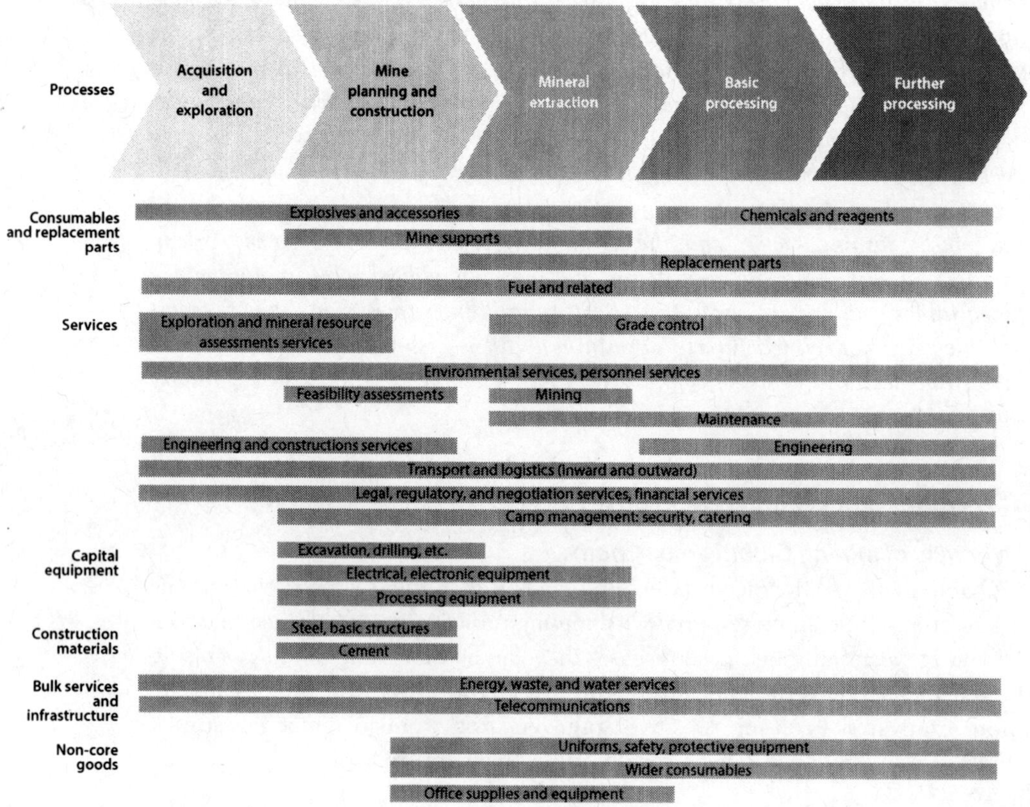

Processes	Acquisition and exploration	Mine planning and construction	Mineral extraction	Basic processing	Further processing
Consumables and replacement parts		Explosives and accessories		Chemicals and reagents	
		Mine supports			
			Replacement parts		
		Fuel and related			
Services	Exploration and mineral resource assessments services		Grade control		
		Environmental services, personnel services			
	Feasibility assessments	Mining			
			Maintenance		
	Engineering and constructions services		Engineering		
	Transport and logistics (inward and outward)				
	Legal, regulatory, and negotiation services, financial services				
	Camp management: security, catering				
Capital equipment		Excavation, drilling, etc.			
		Electrical, electronic equipment			
		Processing equipment			
Construction materials		Steel, basic structures			
		Cement			
Bulk services and infrastructure	Energy, waste, and water services				
	Telecommunications				
Non-core goods		Uniforms, safety, protective equipment			
		Wider consumables			
		Office supplies and equipment			

Source: World Bank 2012.

Value Chain Dynamics

There are a number of trends that affect the value chain structures and dynamics, including the following:

- *High commodity prices* are fuelling a mining exploration boom. Global exploration spending reached US$18.2 billion for nonferrous metals in 2011. Iron ore exploration for 2011 was estimated at US$2.5 billion. This is leading to expanded opportunities and a widening spatial coverage of FDI in the sector. Overall, high commodity prices have strengthened the competitive position of major mining companies in the value chain. The additional profits have allowed firms to acquire junior mining companies and conduct further in-house exploration activities.

- *Emergence of nontraditional investors*, especially from the BRICS,[2] is particularly high in Sub-Saharan Africa. Most prominent here is China, whose mining investment in Africa reached US$15.6 billion in 2011 across 16 countries. Chinese investment in the mining industry is usually through joint ventures with local firms or the direct purchase of equity of local entities (Mlachila and Takebe 2011). This has implications for the mining value chain, and the supply chain in particular, as evidence from Zambia suggests that Chinese mines have a different approach to their supply chain—encouraging open access to the supply chain, but not supporting supplier development (Morris and Fessehaie 2012).

- *Technology improvements* are also allowing more efficient mining techniques that maximize returns and minimize extraction and processing costs. Technology improvements have an impact on value chain dynamics as the imperative for cost-efficient mining techniques encourages cooperation between mining companies and their suppliers to improve operational efficiency.

Value Chain Governance

The mining value chain is producer driven, with the competitive advantage of major mining companies lying in production know-how rather than product design or marketing. Major mining companies are the lead firms in the sector and set standards and performance parameters for value chain participants. The position of mining companies as lead firms has been strengthened by high commodity prices, which has increased mining companies' margins.

In addition to mining companies, governments are developing as key actors in the mining value chain. The particular time-bound nature of mining (with a defined mine lifespan, although in some cases this can be quite long term) creates a motivation from resource-rich countries to maximize socioeconomic benefits and to reduce dependency on mining value chains in the long term. In particular, countries target infrastructure development, staff localization, and local procurement. Communities too are increasingly seen as providing mines with a "social license to operate," with mines responding by emphasizing sustainability issues.

As an example of the direction of this new deal-making, a recent McKinsey report estimated that approximately a quarter of FDI resource investment deals over US$250 million into Africa in between 2006 and 2010 included investment in infrastructure and/or resource processing, as opposed to only 1 percent of deals in the 1990s.

For mine development, mining firms tend to contract out operations with firms specializing in engineering, procurement, and construction management (EPCM) as primary contractors. EPCMs in turn subcontract to specialized goods and services providers, including original equipment manufacturers (OEMs). Depending on the agreement, mines may have some influence over which subcontractors are hired by the primary contractors. This reflects a modular value chain governance type. OEMs may in turn procure a range of components and services from lower tiers of specialized suppliers—indeed, EPCMs often form strategic partnerships and joint alliances to cover a larger scope of infrastructure required.

Mining companies tend also to have direct established relationships with OEMs, where the capital equipment and the associated training, repairs, and maintenance may be supplied through a distributor (for example, Barloworld, or Manutention in the case of Caterpillar), or through a subsidiary (for example, Sandvik), depending on the company's strategy in the particular geography. Key suppliers can also exercise some power in value chains through specialized technology, exclusive relationships, or highly concentrated production activities. For other strategic items such as chemicals, reagents, and fuel, mines also tend to negotiate a direct agreement with the supplier—often through centralized procurement operations. Other inputs (in particular noncore goods and services) may be managed by intermediary inventory management firms/services firms, which reduces the number of purchase orders issued by the mining companies and lowers transaction costs. For example, the French facilities management firm Sodexho provides turnkey camps and an integrated hospitality service to mining companies, sourcing catering and laundry services and security providers as part of its contract.

These relationships have implications for local suppliers, depending on the type of product segment targeted. In particular, modular and relational governance observed in the mining value chain will mean that smaller local suppliers (particularly in noncore goods) may need to engage with "Tier 1" suppliers (such as EPCMs, camp management, inventory management suppliers) to enter the mining value chain. Stakeholders involved in mining procurement indicated a preference for employing service companies in their supply chain to manage the high number of smaller suppliers, in order to directly reduce the transaction costs and coordination efforts of the mining company.

Mining Activities in Ghana, Mozambique, and Chile

Ghana is Africa's second-largest gold producer after South Africa, producing approximately 4 percent of the world's gold. Gold mining activities extend across ore extraction (both open pit and underground mining) and processing (including carbon-in-pulp, carbon-in-leach, bio-oxidation, and heap leaching processes), as well as exploration and mine development activities. Ghana is also a major producer of manganese, which is its largest mineral product by volume. Other commodities, including uranium, iron ore, and diamonds, offer prospects for development. In terms of beneficiation, the government of Ghana has prioritized the development of an integrated aluminum industry, and has restarted production at its Valco aluminum smelter in Tema.[3] Investment in a gold refinery has also been targeted by the Modern Gold Refinery, part of the Chyuan Chya Group, and other investment partners.

Mozambique's mineral industry is, at present, dominated by tantalum and ilmenite mining. However, the country is highly likely to yield substantial coal deposits and coal exports are expected to grow dramatically in the coming years. Other commodities include gemstones and limestone, and there are known reserves of pegmatite, platinoids, uranium, bentonite, iron, cobalt, chromium, nickel, copper, granite, fluorite, diatomite, emeralds, tourmaline, and apatite in the country. In addition to mining activities, processing is occurring at BHP Billiton's aluminum smelter, Mozal, which is the second-largest smelter in Africa. At present, aluminum accounts for approximately 60 percent of Mozambique's exports.

Chile is one of the world's biggest producers across a range of minerals—particularly nonferrous metals such as copper and molybdenum. It also has substantial gold and silver deposits and declining iron ore deposits, as well as large deposits of industrial and energy minerals such as iodine. Codelco, the autonomous state-owned corporation, dominates copper production as the largest copper producer in the world. There are also approximately 1,500 small-scale producers in Chile, and another state-owned business, Empresa Nacional de Mineria (ENAMI), that promotes small- and medium-size private sector mining in Chile by purchasing copper from approximately 450 small-scale mines for further processing at ENAMI-owned processing facilities. Copper concentration activity is undertaken on-site or at company-owned concentration plants in the vicinity, and some mining companies also undertake copper cathode production on-site. State-owned mining companies account for 80 percent of the refinery and smelting operations in the country.

FDI in the Mining Sector in Ghana, Mozambique, and Chile
Overview

Foreign investment in the mining industry within each country varies in terms of scale, commodity, and nationality of foreign investor. In **Ghana**, mining production has resulted from investment in existing gold mines rather than greenfield projects. A number of major global mining houses are operating in Ghana, including Anglo-Gold, Goldfields, and Newmont. Investment in manganese and bauxite has been driven by vertical integration of companies involved in downstream production of manganese alloys and aluminum, and includes presence of non-Western investors, including Privat Group (Ukraine) and Bonsai Minerals (China).

In **Mozambique**, foreign investment has been in greenfield operations due to the nascent development of the mining sector. Top-tier mining companies such as Rio Tinto and Vale SA have made investments in the coal sector in recent years.

But smaller companies are also involved, such as Kenmare Resources, which produces heavy minerals such as ilmenite, rutile, and zircon. Significant greenfield exploration activity is under way in Mozambique, particularly in Tete Province (Baobab Resources of Australia and Eurasian Natural Resources Company of Kazakhstan) and Manica Province (Pan African Resources of South Africa and Great Basin Resources of Canada).

In **Chile**, foreign investment in the mining industry has been driven by acquisitions of existing mine sites by foreign investors, with only limited greenfield investments. Many of the major international mining houses are present in the country, including Anglo-American, BHP Billiton, Barrick Gold, X-Strata, and Kinross. These investors mainly operate in consortia of joint ventures across foreign firms, which is a key feature of FDI in the Chilean mining industry.

Drivers of FDI Location Decisions

Key drivers of FDI indicated by survey respondents across the selected countries (see table 5.1) included political and social stability, an environment conducive to business, and access to raw materials.[4] In Chile, investors also see the country as a good base for operations within the region (high importance given to access to local and regional markets). In line with these concerns, proximity to raw materials was ranked as most important in terms of geographical location of investment. In Chile firms also indicated proximity to suppliers as an important factor in locating investment (perhaps due to the presence of a greater number of local suppliers). In contrast, this was not as important in Mozambique, where there are fewer local suppliers and firms would tend to have a greater reliance on supply from South Africa.

Table 5.1 Drivers of Investment by Foreign Mining Companies

	Overall ranking	Ghana	Mozambique	Chile
Political and social stability	1	1	2	1
Access to raw materials	2	3	1	6
Business environment	3	2	7	2
Access to land/facilities	4	3	3	6
Access to specific inputs	5	6	3	4
Fiscal incentives	6	3	3	12
Preferential market access	7	12	3	4
Access to local (and regional) markets	8	11	8	3
Access to reduced labor costs	9	7	9	12
Access to skills	9	7	12	9
Personal relationships	9	13	9	6
Following our competitors	12	13	9	9
Access to reduced nonlabor costs	13	7	12	15
Access to technology	14	7	14	14
Following our multinational customers	15	13	15	9
Other	16	16	15	16

Source: Surveys of 14 foreign-owned mining investors: Ghana = 4, Mozambique = 5, Chile = 5.

Mining legislation is also critical for setting the investment context in each country. Mining legislation outlines the rules of engagement for foreign investors, including any limitations on repatriation of profits, time frames for utilization of exploration licenses, and any additional regulations in terms of mining-specific incentives, localization, and local procurement. These are discussed in detail in the later section "Factors Contributing to Spillovers."

Overview of Domestic Industry in the Mining Value Chain

Within the selected countries, mining supply industries have developed to different extents (see figure 5.2a). Ghana's mining supply industry includes investment by foreign suppliers, as well as domestically owned suppliers that have emerged over the long period during which mining has taken place in Ghana. Mozambique's supply industry is much less developed. Chile is seen as having a well-developed mining supply sector that has grown to supply not only mines in Chile but also those elsewhere in the region.

Domestically Owned Suppliers

In **Ghana**, the development of a domestically owned supplier base is a process that has taken many years. Active programs by government, the mines, and multilaterals have played an important part, as has the overall business operating environment in Ghana. This has resulted in capacity and companies across a number of different areas. These include:

- Civil works, construction, engineering, and consulting services (for example, Engineers and Planners, Electrofax, Naachia)
- Casting and forging of steel-grinding media and mill liners (for example, Tema Steel, Western Castings, Western Forging); however, limitations of equipment, processes, and quality of inputs have constrained these companies' ability to produce at a sufficient quality to meet mines' demands
- Transport and logistics (for example, Allship, Holman Petroleum)
- Plastic piping (for example, Interplast)
- Kilns and furnaces (for example, Caesar Kilns and Furnaces)
- Repairs to hydraulic equipment (for example, Harlequin)
- Crucible and cupel production (for example, Sapat, Nyarko-Mensah)

The above companies have evolved in a variety of different ways. Few of these companies preexisted the mining sector in Ghana, and those that did have significantly modified their operations in response to the mining sector's needs. Some companies were established specifically to serve the mining sector (for example, Caesar, Sapat, Holman Petroleum), and a number of these were created by ex–mine employees.

In **Mozambique,** local participation in the mining supply chain has been limited to service providers, including, for example, engineering, civil works, environmental assessments, laboratory services, spare parts provision, maintenance and repair, transport and logistics, and catering; and input suppliers, such as explosives and building materials. At present, there is limited production capacity in the country.

In **Chile,** local service providers exist in the areas of engineering, consulting, and construction. However, the increasing requirement for integrated services means that supply linkages are generated through large, international contractors such as Aker Kvaerner and Fluor Daniel, which subcontract specific tasks to local firms (Cochilco 2006, 149). Domestic producers often compete against international suppliers that have operations in the country or local distributors of these imported goods. There are limited supply linkages in the maintenance sector, as international firms such as Komatsu supply their goods under "Full Marc" agreements (maintenance and repair contracts). The supply of heavy mining equipment is still dominated by international companies, with the exception of DrillCo Group and Implemin, who produce drilling trucks and pneumatic drills (Cochilco 2006, 149). Chilean producers typically compete in lower-value goods such as conveyors and fans.

Foreign-Owned Suppliers

For foreign-owned suppliers, a similar pattern is found across the countries, with foreign suppliers in Chile operating on much larger scale than in the other countries (figure 5.2b).

In **Ghana**, foreign investment by global mining suppliers and primary contractors accounts for a large proportion of mining supply activity. Presence of international project engineering companies and EPCMs (for example, Lycopodium, Knight Piesold, AMEC Minproc), capital equipment OEM companies (Atlas Copco, Boart Longyear, Sandvik, Liebherr, Mantrac), mining and drilling contractors (Moolmans), laboratories (SGS, ALS), explosives suppliers (Maxam, African Explosives), and cement (Ghacem – Heidelberg Cement Group) is particularly prominent.

Foreign investment by international project engineering companies, OEM companies, and drilling contractors is also in its early stages in **Mozambique**. Liebherr (2011) and Sandvik (2011, 20) both established subsidiaries in Mozambique in 2011, while Atlas Copco established a customer service center in Mozambique in 2012. Other primary contractors, such as Fluor Daniels and Boart Longyear, are also active in Mozambique through their South African subsidiaries. The proximity to the well-established mining services and manufacturing industry in South Africa is likely to limit investment in manufacturing and value-added services in Mozambique. Furthermore, foreign suppliers based in Mozambique must contend with large distances from the commercial center (Maputo) to the mine operations in the provinces. Foreign suppliers indicated an average distance of 1,043 kilometers from headquarters to mine operations, compared to 208 kilometers in Ghana and 691 kilometers in Chile.

In **Chile**, many international suppliers and subcontractors have invested in establishing a presence. Large companies such as Sandvik, Komatsu, Atlas Copco, and Boart Longyear have multiple branches throughout the country offering sales, maintenance, and production services.

Figure 5.2 Average Annual Sales of Domestic- and Foreign-Owned Suppliers

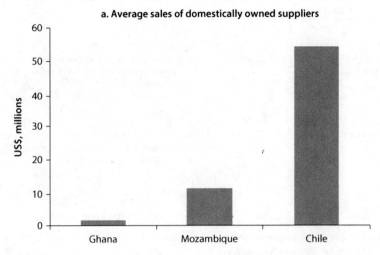

Source: Surveys of 46 domestically owned suppliers: Ghana = 7, Mozambique = 21, Chile = 18.

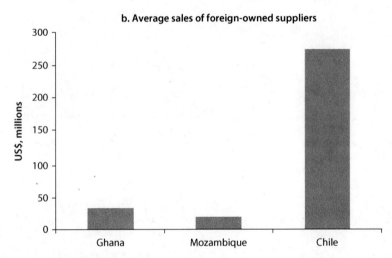

Source: Surveys of 36 foreign-owned suppliers: Ghana = 14, Mozambique = 11, Chile = 11.

Supply Chain Effects

Evidence of Spillovers

Linkages

Drawing on available information from sustainability reports, table 5.2 summarizes mining companies' definitions of "local" procurement/linkages and the extent of local procurement spending. Varying definitions of local procurement across mines make it difficult to draw comparisons. For example, Teck Cominco's low percentage of local procurement spending in Chile can be attributed to the "regional" definition used in measuring local procurement, whereas Gold Fields

Table 5.2 Selected Mining Companies in Ghana, Mozambique, and Chile: Local Expenditures and "Local Procurement" Definition

Company	Definition of "local" for procurement/linkages	Local procurement spending
Ghana		

AngloGold Ashanti[a]	• National
	• Includes indirect imports and locally produced goods
		• Local procurement (2011):
		  – Obuasi: US$280 million (61 percent)
		  – Iduapriem: US$107 million (57 percent)
		• Total expenditure:
		  – Obuasi: US$459 million
		  – Iduapriem: US$188 million

Gold Fields[b]	• No definition provided	• Local procurement: unknown • Total expenditure (Tarkwa and Damang): R5,949 million (US$720 million[j])[k]
Mozambique		
Vale SA[c]	• National and regional definition	• Over US$1 billion in recent years[l]
Rio Tinto[d]	• Four definitions of local procurement: national, regional, local, and community	• Unknown
Kenmare Resources[e]	• Mozambican-registered companies	• Local procurement: 22 percent • Total expenditure: unknown
Chile		
Teck Cominco[f]	• Regional level • No consideration of value added or indirect imports	• Quebrada Blanca: 9 percent • Carmen de Andacollo: 5 percent • Total expenditure: unknown
Minera Collahuasi[g]	• Regional—provides countrywide regional breakdown • No consideration of value added or indirect imports • Excludes locally procured services	• Total expenditure (goods and services): US$1.7 billion • Total expenditure (goods): US$354 million • Santiago Metropolitan Region: US$237 million • Tarapaca Region: US$64.7 million • Antofagasta Region: US$40.9 million
Barrick Gold Corporation[h]	• National • Excludes purchases from outside the country	• Local procurement: Zaldivar mine: US$451 million (2011) • Total expenditure: unknown

Xstrata (Lomas Bayas)[i]	• Regional
	• No consideration of value added or indirect imports
		• Local procurement (2010):
		  – Goods and inputs: US$139.7 million
		  – Services: US$36 million
		• Total expenditure:
		  – Goods: US$207.8 million
		  – Services: US$45.8 million

Sources: a. AngloGold Ashanti 2012, 5; b. Gold Fields 2010; c. Vale SA 2012a, 108; d. Rio Tinto 2012; e. Survey responses; f. Teck Cominco 2012; g. Dona Ines de Collahuasi 2012, 32; h. Barrick Gold Corporation 2011, 129; i. Lomas Bayas 2011, 14. j. Based on exchange rate of US$1 = 8.2612 ZAR; accessed Sept. 21, 2012. k. Likely to be Ghanaian total procurement spend, as not labeled explicitly as "local procurement." l. Campbell 2012.

Sector Case Study: Mining

indicates total procurement spending in Ghana as US$720 million. Furthermore, none of the mining companies highlighted above distinguish between indirect imports (imported goods bought through local agents or regional and national sales offices) and local value-adding companies that manufacture products in-country.

Survey results show that foreign mines operating in Chile have the greatest proportion of domestically owned suppliers (figure 5.3). Furthermore, these firms appear to be carrying out more value-added activities in-country than in Ghana, and particularly more than in Mozambique. These results also show a very low level of foreign-owned suppliers operating in Chile—what Chilean mines source from local companies is predominantly from locally owned companies. This is quite different from Africa, where overall a much higher proportion of domestic firms are foreign owned.

Actual FDI Spillovers through Supply Chains

Assistance by Foreign Investors. One of the key interactions that mining companies have with suppliers is through the support that they provide. Table 5.3 indicates

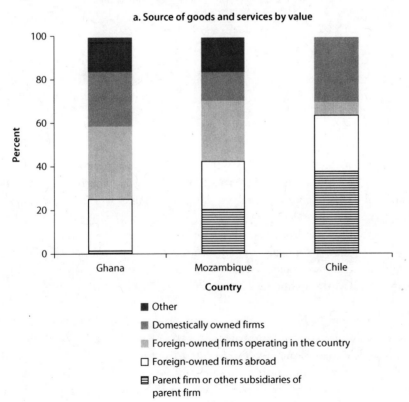

Figure 5.3 Sourcing of Goods and Services by Foreign-Owned Mining Firms

a. Source of goods and services by value

- Other
- Domestically owned firms
- Foreign-owned firms operating in the country
- Foreign-owned firms abroad
- Parent firm or other subsidiaries of parent firm

Source: Surveys of 14 foreign-owned mining firms: Ghana = 4, Mozambique = 5, Chile = 5.

figure continues next page

Figure 5.3 Sourcing of Goods and Services by Foreign-Owned Mining Firms *(continued)*

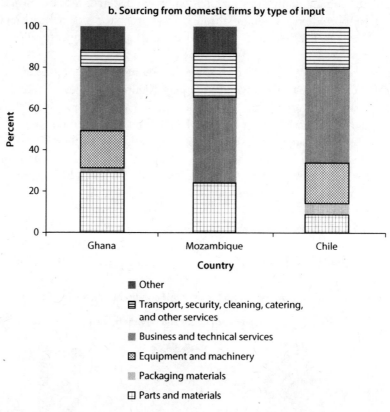

b. Sourcing from domestic firms by type of input

Source: Surveys of 12 foreign-owned mines: Ghana = 4, Mozambique = 3, Chile = 5.

that as part of their relationships with their suppliers, mining companies most frequently provide assistance in terms of implementing health, safety, and environmental (HSE), and/or social conditions; training workers; and providing advance payment.

However, the data from domestic suppliers indicates that this support is provided infrequently across all countries (table 5.4). For example, the ranking for "help with quality assurance" (where 1 = never and 4 = always) was very low for Ghana (1.7), Mozambique (1.7), and Chile (1.2).

Demand by Foreign Investors. Survey responses indicated that the majority of domestic suppliers in all countries (72 percent in Chile, 57 percent each in Ghana and Mozambique) were requested to make specific improvements to their operations before or after signing contracts with foreign firms. The most common improvements required by foreign investors were improved timeliness of delivery; improvements in HSE and social conditions; and investments in new technology and equipment. The general pattern was the same across countries.

Table 5.3 Foreign Mining Firms' Assistance to Suppliers, Ranked by Frequency of Support

Type of support	Overall rank	Ghana	Mozambique	Chile
Help with implementing health, safety, environmental, and/or social conditions	1	1	1	1
Training of workers	2	2	3	2
Advance payment	3	3	1	5
Help with quality assurance	4	4	7	3
Support for sourcing raw materials	5	4	5	7
Help with inventory control	6	4	13	3
Lending/leasing of machines or equipment	7=	8	7	7
Help with audits	7=	8	7	7
Repair/maintenance of machines	9	8	4	11
Product or process technologies	10	8	7	11
Support to get funds from other sources	11	8	7	17
Licensing of patented technology	12=	4	13	17
Provision of financing for improvements	12=	15	5	14
Provision of inputs	14=	8	13	14
Financial planning	14=	14	7	14
Help with finding export opportunities	14=	15	13	7
Help with organization of production lines	17	17	13	6
Help with business strategy	18	19	13	11
Other	19	17	13	19

Source: Surveys of 13 foreign-owned mining investors: Ghana = 4, Mozambique = 4, Chile = 5.
Note: Where a number is shown with an equal sign (e.g. "14="), this indicates that two or more types of support are ranked equally.

Upgrading Technical and Business Capacity, HSE, and Quality. Supplying FDI may also facilitate investment in quality certifications. While only two of seven Ghanaian firms surveyed acquired quality certification, both did so at the behest of FDI customers (figure 5.4). In Chile too, a significant proportion of firms that obtained quality certifications did so to meet requirements of their foreign-owned customers.

Poor HSE practices and limited compliance with standards among suppliers is seen by mining companies as a key barrier to increasing local sourcing—this was particularly noted in Ghana. In some cases, domestic companies have improved their HSE practices and expertise, as well as acquiring standards and certifications (see box 5.1). These may be demanded as part of bidding processes that require HSE expertise and systems, as well as through on-site work where employees need to comply with mines' HSE standards.

Finally, a significant number of enterprises indicated that their main product supplied to mining companies was developed specifically for that client. As part of this product development process (particularly if done jointly with mining companies), companies often needed to build their technical capacity, including upgrading equipment and quality of inputs, putting in place quality control systems, and ensuring adequate utilities such as energy supply. In Ghana, for example, a number of mining companies have worked with Ghanaian-based

Table 5.4 Assistance Received by Domestically Owned Suppliers from Foreign Customers, Ranked by Frequency of Support

Type of support	Overall rank	Ghana	Mozambique	Chile
Help with quality assurance	1	1	3	3
Help with implementing health, safety, environmental, and/or social conditions	2	5	1	3
Training of workers	3	3	2	5
Advance payment	4	1	3	8
Licensing of patented technology	5	3	10	1
Provision of inputs	6=	5	17	2
Provision of financing for improvements	6=	9	10	5
Repair/maintenance of machines	6=	11	5	8
Help with audits	9=	5	7	14
Financial planning	9=	11	7	8
Support for sourcing raw materials	9=	11	7	8
Help with business strategy	12	11	5	12
Help with organization of production lines	13	5	10	14
Product or process technologies	14=	9	10	14
Lending/leasing of machines or equipment	14=	11	10	12
Help with finding export opportunities	14=	11	17	5
Support to get funds from other sources	17=	11	10	14
Help with inventory control	17=	11	10	14
Other	19	11	19	14

Source: Surveys of 45 domestically owned suppliers: Ghana = 7, Mozambique = 20, Chile = 18.
Note: Where a number is shown with an equal sign (e.g. "14="), this indicates that two or more types of support are ranked equally.

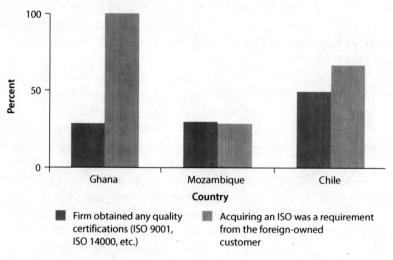

Figure 5.4 Domestic Firms Obtaining Quality Certifications

- Firm obtained any quality certifications (ISO 9001, ISO 14000, etc.)
- Acquiring an ISO was a requirement from the foreign-owned customer

Source: Surveys of 46 domestically owned suppliers: Ghana = 7, Mozambique = 21, Chile = 18.

Box 5.1 Mozambique Mining Supplier Development—Soradio

Soradio is a domestically owned small and medium enterprise electrical contractor based in Maputo. It supplies foreign investors in the mining and agricultural value chains, as well as other sectors, in Mozambique. The owner-manager started the company after having gained experience doing contract construction and electrical work in the Southern African Development Community region for various large companies. This experience helped to build a network and reputation, as well as enhancing understanding of the requirements of FDI, which were recognized as being different from those of local firms and the public sector market. Learning English was also important for working more easily with companies elsewhere in the region, particularly foreign companies and suppliers that do not speak Portuguese.

The company was one of the first to work with Mozal (prior to the MozLink program), and has experienced a significant learning curve since then. Soradio made significant efforts to become certified and to comply with quality, health, and safety requirements; this was a substantial investment not yet required by other clients in the market. Personal reputation and networks played an important role in building the company capacity and profile over time. Relationships with primary contractors were particularly important elements of success. For example, safety officers with primary contractors helped Soradio to understand what it would take to comply with mines' requirements. Furthermore, Soradio was not familiar with tendering processes, and had to learn how to price projects, complete bid documentation correctly, and comply with specifications (such as uniforms, insurances, and bid bonds). Much of the learning process happened through observing and personal interaction.

Having worked successfully with one FDI company and built a good reputation, Soradio has found it easier to win work with other FDI companies. However, the scale of Soradio's growth has been limited by lack of finance. Expansions can only be funded internally or through its networks, as local banks have limited offerings for small businesses and are not an affordable option.

Tema Steel and Western Castings to improve the quality of local steel ball supply. Support provided included testing of products or sharing of testing costs (both in Ghana as well as conducting laboratory tests in South Africa) and advisory on required raw materials and processes by company metallurgists. In Chile, a number of collaborations in product development have taken place as part of the World Class Supplier Program, including projects to expand cable shelf life, to improve air quality, and extend tire life (Urzúa 2011).

Growth Opportunities. Survey results indicate that the value supplied by enterprises to the mining sector has climbed significantly in Ghana and Chile (and to a lesser extent in Mozambique) since enterprises began supplying mines. This increase in value is as a result of enterprises being able to access larger contracts, progress to supplying higher-value goods, or expanding their customer base. Enterprises indicated that supply to an initial foreign mining company was also instrumental in gaining access to other opportunities (see figure 5.5). Clearly the

Figure 5.5 Value of Domestic Firms' Output to Foreign Customers and Market Links

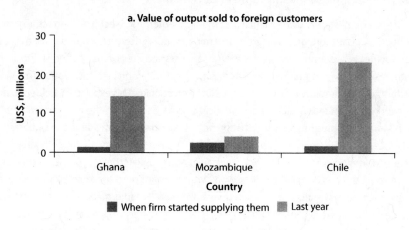

a. Value of output sold to foreign customers

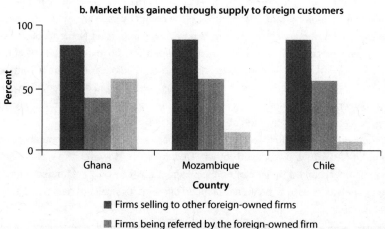

b. Market links gained through supply to foreign customers

■ Firms selling to other foreign-owned firms
■ Firms being referred by the foreign-owned firm with whom supplier first started
■ Firms selling to foreign-owned firms affiliated with the foreign-owned firm with whom supplier first started

Source: Surveys based on 46 domestically owned suppliers: Ghana = 7, Mozambique = 21, Chile = 18.

far less significant increase in Mozambique is due to the mining sector being in the early stages of development. Ghanaian and Chilean firms' sales to foreign-owned customers have both increased, on average, by a factor of approximately 12. Under the right conditions, Mozambique may also achieve this in time.

The majority of suppliers surveyed across the three countries indicated that they supplied more than one mining company, often having been referred by their initial foreign customer. In Ghana, a number of suppliers are planning to expand capacity or develop new capacity to supply the mining sector. Tema Steel (a foreign-owned supplier) is exploring the potential to double its capacity to produce grinding media for the mining sector. Kofi Ababio & Sons, an importer and distributor of chemicals for mining, is exploring the possibility of moving up the

Box 5.2 Supplier Upgrading in Chile—Drillco Tools

Drillco Tools is a Chilean-owned designer and manufacturer of drilling equipment for the mining, construction, and energy industries, with a particular focus on drill bits for open pit mining and exploration. Since the company's formation in 1990 as an importer of drilling equipment for the mining industry in Chile, it has upgraded its capabilities and diversified both its products and markets to become a world-class manufacturer and innovator in drill hammers.

Figure B5.2.1 Evolution of Drillco (1990–2012)

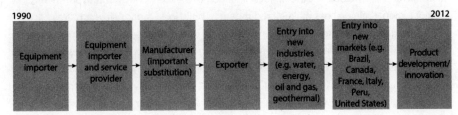

Today, the company employs 170 full time employees and exports 70 percent of its products through its network of distributors to a diverse set of markets, including Canada, Brazil, France, Peru, Poland, and the United States. Drillco competes successfully against foreign suppliers including Atlas Copco and Sandvik.

Drillco Tools is also developing new, innovative products in house for the mining industry. One example is the Sniper RC5 hammer, a patented drill bit for use in exploration drilling.

Source: AreaMinera 2010, 23.

---

value chain through the construction of a 20,000-ton caustic soda plant in Ghana. A number of companies have developed local manufacturing capacity after supplying to mining companies—an example from Chile is outlined in box 5.2.

Finally, a number of firms have started exporting as a result of supplying foreign mining companies. Of nine domestic suppliers surveyed in Chile, four began exporting directly as a result of their supply to FDI. In Ghana two of the six firms surveyed had become exporters.

Factors Contributing to Spillovers
Global Value Chains and FDI Characteristics

Despite the shift toward centralized procurement and supply chain management in mining multinationals, survey results indicate that a significant share of procurement decisions is made by domestic management across all countries—88 percent of the value of goods determined by local management in Ghana, 78 percent in Chile, and 69 percent in Mozambique. The lower levels in Mozambique may also reflect its location near South Africa, where a number of mining firms such as Rio Tinto have strategic operational hubs (see box 5.3).

This continued role for local decision making seems to reflect that mining companies increasingly see their support in developing a local supply base as

Box 5.3 Does Location Matter?

Proximity to regional offices and countries with established mining industries can mean that the important decisions of foreign-owned mining companies and foreign-owned suppliers are taken outside of the country. For example, BHP Billiton's human resource, finance, and procurement services for their mining operations in Peru are managed out of their office in Santiago, Chile. Decisions of foreign-owned mine engineering firms in Peru are also taken in company head offices based in either Chile or the United States. This dynamic is clearly also affecting the potential for Mozambique to develop local linkages, as the role of neighboring South Africa as a major regional mining base makes it redundant to build local capacity in many areas.

For Ghana, location presents an opportunity. Its geographic position offers suppliers an attractive hub within West Africa for serving the region. One supplier that uses Ghana as a regional hub is Geodrill, a West African exploration drilling company. It supplies drilling operations to mining investors in Ghana and Burkina Faso, and has plans to enter Niger and Côte d'Ivoire. Another supplier with a hub in Ghana is West Africa Tyre Services Ltd. This Accra-based company provides tires and tire management services to the mining industry in Burkina Faso, Guinea, and Mali.

Source: Fernandez-Stark, Bamber, and Gereffi. 2010.

---

critical in gaining a "social license to operate." Along with these social obligations, mining companies favor and support the availability of inputs locally where this contributes to achieving key supply chain objectives, including ensuring that inputs are delivered at the required quality standard, to the required scale, and at the correct time.

The existence of supply relationships is the starting point for potential supply chain spillovers. However, achieving spillovers depends in part on the nature of the interaction between foreign investors and local suppliers. Figure 5.6 highlights that formal contractual relationships between suppliers and mining companies—particularly contracts of more than a year—are most common in Chile, perhaps reflecting the more established relationships in the mining industry in the country. The length of contract matters because it appears to influence the likelihood that foreign investors provide assistance to supporting capacity building of local suppliers (see later discussion).

Domestic Supplier Characteristics

Table 5.5 summarizes the main obstacles preventing sourcing from domestic firms across the survey countries, as identified by foreign investors. A greater proportion of foreign investors in Mozambique indicated that they face difficulties in sourcing from local suppliers (43 percent) than in Ghana and Chile (14 percent and 11 percent). Mining companies are able to procure higher-value and more sophisticated inputs from domestic companies in Mozambique than in Sub-Saharan African countries such as Ghana and Mozambique.

Sector Case Study: Mining

Figure 5.6 Domestic Firms' Sales to Foreign Firms by Type

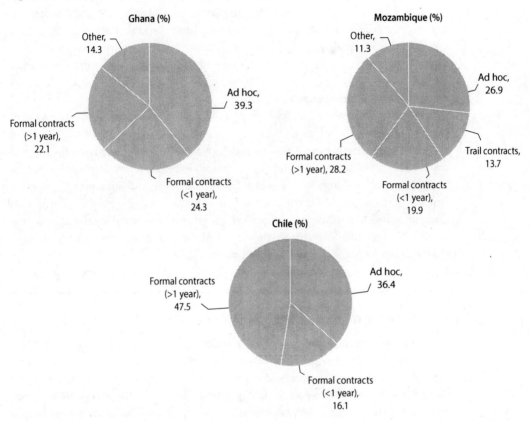

Source: Surveys of 40 domestically owned suppliers: Ghana=7, Mozambique = 15, Chile = 18.

Table 5.5 Obstacles to Sourcing from Domestic Firms

Factor	Overall ranking	Ghana	Mozambique	Chile
Inadequate quality control system	1	2	1	1
Lack of suitably trained and skilled personnel	2	4	3	2
Production capacity is too small	3=	1	8	2
Unable to make timely deliveries	3=	2	5	4
Poor engineering and design capabilities	5	4	3	6
Pricing is uncompetitive	6	8	2	6
Unable to finance investment necessary to supply mining firm	7=	6	5	6
Low level of technology	7=	6	5	6
Do not have quality certificates required by suppliers	9	10	10	4
Cannot meet health, safety, environmental, and/or social standards	10	9	9	10
Other	11	10	10	11

Source: Surveys of 14 foreign-owned mining firms: Ghana = 4, Mozambique = 5, Chile = 5.
Note: Where a number is shown with an equal sign (e.g., "7="), this indicates that two or more types of support are ranked equally.

In Ghana in particular, poor quality and HSE practices of domestic firms were raised as main barriers to expanding and diversifying supply by domestic firms to mining companies. Limitations due to inadequate production equipment, processes, and poor quality of input materials constrain the ability of domestic suppliers to achieve production at a sufficiently high quality to meet mines' demands.

Host Country Characteristics

In terms of *policy and regulation* specifically aimed at expanding local procurement, a number of options may be implemented to support the development of a local mining supply base. These include set-asides and targets for local sourcing; procurement processes that give preference to local suppliers; and application of taxes and duties on imports—see box 5.4 for an example from Australia. Foreign firms in Ghana and Mozambique that participated in the survey indicated that they are subject to local procurement requirements to source locally and to promote linkages with local firms, although only in Ghana did firms indicate that they needed to source a specific percentage locally.

Box 5.4 Australian Industry Participation Plans

The Australian Industry Participation (AIP) National Framework—established in 2001—supports supply chain development in the mining industry by requiring Australia industries to be given "full, fair and reasonable access" to opportunities deriving from significant public and private investment projects (including mining operations). AIP policy obliges investment projects to compile an AIP plan to set out how they will provide full, fair, and reasonable access to opportunities for Australian suppliers.

While not mandatory for operating, an AIP plan—approved by AusTrade—is a requirement for firms that wish to apply for tariff concessions on eligible imported goods under the Enhanced Project By-law Scheme. AIP plans must show evidence of consultation with Australian manufacturers to determine existing local capacity. Criteria included in the assessment of an AIP plan include, among others: employment creation; skills transfer; regional economic development; technology transfer and R&D; and "full, fair and reasonable opportunities" for suppliers to tenders.

In July 2012, reforms to the AIP program were implemented. They included a requirement for plans and outcomes to be published and a requirement for projects accessing the import duty programs to increase the level of detail provided to AusTrade in their AIP plan. Required details include:

- Demonstration of how AIP commitments cascade down to Tier 1 and engineering, procurement, and construction management companies
- Provision of more comprehensive evidence of opportunities being made available to Australian industry
- Provision of regular reporting on AIP and outcomes for large-scale projects.

Source: Warner 2011, 8.

Sector Case Study: Mining

In **Ghana**, development of regulations (in consultation with private firms) to support local procurement by the mining sector has influenced the emergence of a local supply base in Ghana. Draft regulations developed through this process enact provisions relating to local content of goods and services in the Ghana Minerals and Mining Act (2006). These regulations require mining companies to submit a five-year local procurement plan, for approval by the Minerals Commission, which should include targets and strategies for increasing local procurement (including development of capacity of suppliers). The regulations allow for annual review of the plan by the Minerals Commission. To support achieving targets, the regulations require that when bids are within 2 percent of price, the bid with highest local content shall be selected. The regulations also provide for a "local procurement list" of specified inputs that should be purchased locally.

In **Mozambique**, there are no commitments to local supplier development in the Mining Law (2002) or Mining Regulations (Decree no.62/2006). There may be some requirements included in the revised mining law due to be submitted to ministers by the end of 2012 (Esterhuizen 2011).

In contrast to Ghana and Mozambique, **Chile** has not put in place any policy or regulation requiring procurement locally. Instead, it provides support for the development of a supply base through supplier upgrading programs in collaboration with the private sector. Chilean policies set a strong legal framework to encourage mining concessions and foreign capital investments from the early 1980s. The policy framework protected mining concessions by a Property Rights Guarantee and guaranteed free repatriation of profits and capital, tax stability, and the principle of nondiscrimination against foreign investment (Republic of Chile 1974).

Effective *industry associations* in the mining industry also support supply chain development by providing a forum for collaboration between mining firms on supply chain and procurement issues. In cases where these associations include suppliers and mining firms, these institutions can also support information sharing and enhanced access to opportunities. In the case of Ghana, the Ghana Minerals Commission developed its local procurement regulation through a process of proactive engagement with the private sector through the Chamber of Mines (box 5.5).

Box 5.5 Ghana Chamber of Mines

The Chamber of Mines in Ghana includes an affiliate group of 43 nonmining companies, including suppliers, financial institutions, and service providers that are both domestic and foreign firms. The Chamber has a supply manager subcommittee that meets regularly to discuss supply chain challenges and approaches to supporting local procurement in Ghana. The subcommittee recently coordinated sharing of information to identify 27 product categories that are already manufactured in Ghana, or that may feasibly be manufactured in Ghana within a 5–10 year period. It has collaborated with the International Finance Corporation to conduct a study examining the capacity of local enterprises in Ghana.

The *investment and operating environments* in Ghana and Mozambique are seen as limiting the establishment and growth of local businesses, as well as the development of supply relationships between mines and domestic suppliers. In contrast, a number of stakeholders mentioned the stable, probusiness, and competitive investment environment in Chile as a key driver in the development of its local supply industry and wider economic success. This competitive environment resulted from wide economic reforms implemented between 1973 and 1990, which supported nondiscrimination toward foreign businesses through elimination of trade barriers.

In Sub-Saharan Africa, lack of access to business premises, unreliable supply of utilities, high costs of capital, and difficulties accessing finance are some of the key challenges that make it difficult for entrepreneurs firstly to establish businesses and secondly to compete effectively with foreign-owned suppliers. In Ghana, a high cost base, and, in some cases, unreliable services and utilities (particularly electricity), impede the ability of local enterprises to invest and develop capacity to serve foreign suppliers. In particular, the high cost of capital (for example, interest rates of over 30 percent), high costs of property rental, and high electricity costs were all raised by stakeholders as prohibitive. Also, the limited nature of industrial activity in Ghana means that many inputs and raw materials cannot be sourced locally but must rather be imported. Furthermore, increasing competition with the growing oil and gas sector for services, skills, and facilities has been mentioned as contributing to a rising cost base for local enterprises.

The *available skills base* also affects domestic businesses, and their ability to form and deliver on supply relationships with mining companies. Chile's strong skills base has been highlighted as contributing to the development of a strong local supply base. The government emphasizes tertiary education and spends 1.8 percent of its gross domestic product (GDP) on tertiary education—more than the Organisation for Economic Co-operation and Development average of 1.6 percent. Notably, Chilean engineering degrees are six years in duration, which mining subcontractors in Chile have noted as providing a competitive advantage over the four-year undergraduate programs in the United States, as engineers entering the labor market are more focused on their engineering careers (Fernandez-Stark, Bamber, and Gereffi 2010).

Weak contract enforcement was identified by stakeholders as a key issue that has a negative impact on the development of linkages between foreign and local companies. In Ghana, difficulty enforcing contracts was identified as a key issue by mining companies. This is because the inadequate *legal environment* limits the ability of firms to take legal action against suppliers in case of nondelivery or delay in delivery of mine-critical inputs—making the risk of local sourcing unmanageable. A related point is the actual or perceived lack of transparency in how contracts are often awarded (particularly among state-owned companies). Overall, Chile is regarded as relatively corruption free according to Transparency International's Corruption Perception Index, while Ghana and Mozambique perform substantially worse.[5]

Finally, a complex *regulatory system* may also limit small business and supplier development. In Mozambique, the system is seen as not being geared toward the needs and requirements of small and medium enterprises (SMEs) (UNCTAD 2012, 69). In particular, the complexity of the licensing procedures—which vary depending on factors such as economic activity, location, size, and legal status—can prohibit growth and development. Disincentives to formalization of enterprises, including increased administrative burdens and greater risk of inspections on taxation and labor law compliance (Krause et al. 2008), remain obstacles to the development of a local supply base in the country.

Support for Spillovers
Enhancing Access to Supply Chain Opportunities
Local procurement plans are emerging as tools for increasing local suppliers' access to opportunities. Ghana is currently debating legislation that will require each mining company to produce an annual local procurement plan. At present, Newmont Ghana is the only mining company in the countries under study to have developed a local procurement plan (box 5.6), but both Rio Tinto in Mozambique and Goldfields in Ghana have also commenced work on local procurement policies.

More recently, in response to regulation developed by the Ghana Minerals Commission, the major mining companies have undertaken a collective effort through the Chamber of Mines to identify and support realization of opportunities for local production of inputs or local service provision. In particular, they have started to support development of capabilities in manufacturing, as well as manufacturing of more technical products (for example, crusher liners). The initial stage of this process identified products representing an opportunity for production/delivery in Ghana, based on scale of requirement and perceived ease of production in Ghana.

In Canada and Australia, where there is significant political tension over mining activity on the land of aboriginal communities, community participation agreements are signed prior to mine construction that identifies *contract set-asides* for local suppliers. The contracts are then often not tendered but negotiated with local firms directly. Contract set-asides are also in development in Mozambique at Rio Tinto's proposed coal mine, supported by an in-depth analysis of goods and services with local procurement potential.

Due to the important role that EPCM and primary contractors play in mining operations, some mining companies ensure that local procurement commitments are also written into contracts with their primary contractors (box 5.7). By *incentivizing local procurement at an EPCM level*, mining companies support the broadening of opportunities for local suppliers and SMEs. EPCM local procurement requirements are not a feature of mining agreements in Africa, but these requirements are becoming more prevalent in Canada and Australia.

Additionally, some support programs for mining suppliers provide services across these areas. For example, Rio Tinto's proposed Business Center in Tete

Box 5.6 Newmont Ghana's Local Procurement Policy

The Newmont Ghana mining company ran the successful Ahafo linkages program from 2007 to 2010. This program trained 53 local suppliers in the area immediately surrounding the Ahafo mine, contributing to US$14 million in local procurement. Newmont has more recently rolled out a local procurement policy in Ghana, which outlines areas of support and preferences to be given to various categories of companies (based on geographic location and level of Ghanaian ownership).

Under the local content policy, Newmont aims to increase local expenditure each year, with a higher share going to Ghanaian firms with highest local value added. Among the areas of support provided are:

- Increase Newmont's awareness of goods manufactured in Ghana through:
  - A formal process of supplier registration
  - Identifying products currently being purchased by other mining companies
- Broaden access to opportunities for potential suppliers:
  - Supplier open days
  - Greater use of open tendering, including for all major longer-term service contracts for mining operations as existing agreements expire
  - Advertise available work and goods requirements via the Internet
  - Publish local spending profile data on a quarterly basis
- Apply preferences in assessing tenders in the following order (all else being equal): "local-local" companies; Ghanaian owned; Ghanaian participation; Ghana registered; international
- Build capacity of local companies through the development of collaborative partnerships between industry, nongovernmental organizations, and existing foreign and Ghana-registered companies
- Apply the policy across operational and capital spending
- Effectively communicate policy and achievements, including through supplier open days, the West African Mining and Power Conference, and online reporting on local spending

Source: Newmont Ghana 2010.

---

will be a one-stop shop to support local suppliers with information dissemination, technical support, and training focused on providing information on upcoming opportunities, financing options, prequalification processes, HSE requirements, and procurement and tendering principles. It will also host external specialists to assess local businesses in terms of quality, HSE performance, managerial capabilities, and to provide training and mentoring support in terms of tender preparation and business improvement.

The establishment of *supplier databases* and vendor registration systems also support access to supply chain opportunities in the mining industry. Supplier databases can be publicly or privately managed and are used to prequalify

Box 5.7 Vale SA Supplier Development Program (2012)

The Supplier Development Program was launched in 2012 as a partnership between Vale and the Centro de Promoção de Investimentos. The program targets local SMEs that are operating in Vale's region of influence, providing diagnostic services, firm-specific training, and professional guidance. Eligible firms must demonstrate the potential to provide goods and services to the wider local market (not just Vale), a commitment to the program, and willingness to adapt to good business practices. The project commenced in August 2012 with the registration of companies and by the start of September 2012, 32 companies had registered for the Diagnostic Evaluation stage, of which 17 were preselected for diagnostic evaluation, and another 10 were required to deliver further information. The selected firms are active in areas such as cleaning services, catering, transportation, maintenance, and accommodation.

Vale SA's Supplier Development aims to promote economic development in Mozambican regions under the influence of Vale's mining operations through support to local SME suppliers—Tete, Beira, and Nacala.

Since construction began on the Moatize mining project, Vale has awarded contracts to approximately 813 companies—of which 54 percent were based in Mozambique. Vale also awarded contracts to 179 (22 percent) of firms located in Tete Province directly. These contracts are typically in low-value, noncore goods and services such as maintenance, cleaning services, catering services, and transportation; but some contracts are in areas such as drilling, laboratory services, and tire retreading.

Source: Vale SA 2012b.

---

potential suppliers (for example, registration, tax compliance, capacity, financial analysis) and provide a database of suppliers for mining firms to access easily and reduce transaction costs for mining companies. Examples of two supplier databases are outlined in box 5.8.

Supplier exhibitions and local buyer-seller forums also act as mechanisms for supporting supply chain linkages. In Ghana, the Association of Ghana Industries has held a series of exhibitions focusing on local content, initially focusing on oil and gas local procurement in May 2010, followed by a broader conference and exhibition in December 2011. The association also cooperated with the Chamber of Mines by publicizing the Chamber of Mines' Buyers-Sellers forum held in April 2012. Similarly in Chile, the Antofagasta Industry Association (AIA) hosts an annual mining exhibition—Exponur—which is supported by the foreign mining companies and the government of Chile.

Technical and Business Management Support

There are also a number of programs that offer technical support to local suppliers. Technical support aims to build the capacity of local suppliers to enable local businesses to meet the product performance and quality, HSE standards, scale, and consistency of delivery demanded by mining companies. Support may be

> **Box 5.8 Supplier Databases in Chile and Australia**
>
> **SICEP—Vendor Qualification System (Chile)[a]**
> - Established in 2001 by the Antofagasta Industry Association
> - Qualification system and supplier database for the mining and energy industry in Chile
> - Currently used by 20 purchasing companies, including international suppliers (for example, Komatsu, ABB) and primary contractors (for example, AMEC) and has information on over 2,500 suppliers
> - Qualification activities include:
>   – Registration, evaluation, and qualification of suppliers
>   – Certification of labor and social security compliance
>   – Financial and legal analysis
>   – Training for improving operations
> - Suppliers pay a registration fee and annual maintenance fee for inclusion on the database
>   – Annual cost is between US$200 and US$400
>
> **Industry Capability Network (ICN) Gateway (Australia)[b]**
> - Has been in operation for close to 30 years
> - Covers public and private supply opportunities in a range of industries including oil and gas, energy, mining, water, steel, and transport
> - Approximately 60,000 suppliers and 130 projects on the system
> - More than 100 technical consultants in their network
> - Closely linked to ICN's supplier development activities
> - Publicly funded—free for suppliers to register
>
> *Sources:* a. Stakeholder interviews; http://www.sicep.cl/ingles/archivos/FolletoSicep2009.pdf.
> b. See http://gateway.icn.org.au/.

provided to assist companies enter the mining supply chain, or to upgrade or shift product or service delivery. Programs can also support innovation in the supply chain where there is high competence among local suppliers.

Private Sector–Led Support. Some technical support is conducted in-house by mining companies, such as the MozLink program in Mozambique by the Mozal company. MozLink also combined technical and management support with efforts to improve access to opportunities for local businesses. The program was designed to address the gap between Mozal's world-class standards and those of local SMEs, and the lack of business management skills and experience among local SMEs. Between 2002 and 2007, the program built the capacity of 45 SMEs, providing training in tender preparation and contract execution, and assigning a mentor upon contract award to contract implementation. In this period, Mozal's operational spending with local companies increased from US$5 million to US$17 million per month.

Other technical support is delivered in partnership with third parties, who act as trainers or partners. Examples include:

- *Rio Tinto and AgDevCo multistakeholder supplier development in Mozambique*: In July 2012, a multistakeholder group—led by Rio Tinto and AgDevCo—signed a memorandum of understanding to implement a supplier development program for local sourcing of agricultural produce for consumption by Rio Tinto in Tete province. AgDevCo will manage the project and provide technical and financial support to qualifying companies. Rio Tinto's role in the project will be to provide premises and supply information such as medium-term demand forecasts for food products.

- *Anglo-American Emerge Program in Chile:*[6] Created in 2006, Anglo American's Emerge Program in Chile aims to support SMEs in communities close to its mining operations in country. The program provides entrepreneurship training and technical support on business plan development to successful candidates with businesses achieving sales of between US$400,000 and US$15 million per month. These business plans are then presented to the management committee of the program, which assess the economic viability of the plans and has an option to provide finance and ongoing support to the business for up to three years.

- *Ahafo Linkages Program—Newmont Ghana:* Established in 2007 by Newmont and the International Finance Corporation (IFC), the Ahafo Linkages Program supported local procurement of low-value items such as tools, paints, hospitality services, low-level maintenance and construction, and vehicle rental services. The IFC was responsible for initial business assessment and program design. The IFC also played a role as program coordinators, recruiting implementation partners and identifying groups of SMEs as potential beneficiaries. Newmont Ghana's role involved the creation of a dedicated unit within its supply chain division, which customized procedures for local procurement, educated local suppliers on Newmont's standards, and provided technical support.

Furthermore, in countries where supplier capacity is high and technologically advanced, technical support for local supply chain development can also become a more innovative and collaborative process. In Chile, mining companies are engaging in more innovative supplier development programs that both create value for the mining company and build the capacity of the local supply base. Driven by BHP Billiton, the World Class Supplier Program (box 5.9) focuses on demand-led innovation and close collaboration between the mining company and suppliers to solve challenges faced by mining companies operating in the country.

Public Sector–Led Support. Beginning in the 1990s, the Corporación de Fomento de la Producción de Chile (Chilean Economic Development Agency—CORFO)

Box 5.9 FDI-Driven Innovation and Supplier Upgrading: Chile's World Class Supplier Development Program

The World Class Supplier Development Program reshapes conventional procurement procedures to allow local suppliers an opportunity of offer innovative solutions to challenges faced by mining companies in Chile. Established in 2008 by BHP Billiton, and since expanded to include other mining companies such as Codelco, the project aims to create 250 world-class suppliers in Chile by 2020. The project defines "world class" suppliers as those that are recognized in Chile and abroad, export 30 percent or more of their services, and have innovative and technologically advanced services.

As shown by the diagram below, a review of current Chilean suppliers in the mining industry found that 98 percent of firms were either simple users or adaptors of existing technology and very few were highly innovative in their operations. The aim of the project is to move Chilean suppliers into higher technological areas by 2020. The World Class Suppliers' model encourages mining companies to identify areas where innovative solutions could assist operational efficiency across its operations, and identify local suppliers who have the capacity to work on the problem. Each prioritized challenge is weighted as deemed appropriate and advertised to suppliers. The selection procedure is rigorous: for example, only 16 percent of identified projects at Codelco reached implementation stage. Selection criteria include economic benefit, replicability, urgency of the problem, technological risk, and impact on the operation in the fields of HSE.

Figure B5.9.1 Vision of the World Class Supplier Model

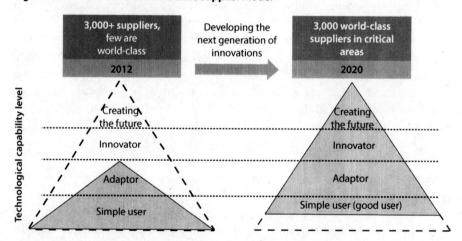

A cluster of 2–3 local suppliers is then created to research the problem and pilot new innovations. In addition to technical funding and support for the area of innovation, some companies such as BHP also employ external consultants to provide training on organization and managerial competence and support supplier linkages with local universities.

The project is coordinated by Fundacion Chile, a nonprofit corporation that aims to support technology transfer and innovation, and increase the competitiveness of Chilean firms

box continues next page

Box 5.9 FDI-Driven Innovation and Supplier Upgrading: Chile's World Class Supplier Development Program *(continued)*

across the economy, but does not provide funding for the projects themselves. Fundacion Chile has also produced a guide for mining companies undertaking World Class Supplier projects and a detailed handbook is also in development. The guide includes practical advice for companies in the operationalization of the project.

As of 2012, 70 cluster projects were under way from BHP and Codelco and there is interest from the Peruvian and Colombian mining industry. BHP is also currently working on ways to clarify intellectual property rights for innovations developed under the cluster program.

Sources: Barnett and Bell 2011; Urzúa 2011.

---

implemented a number of programs that stakeholders highlighted as playing a key role in driving the development of the local supply industry. In 1995, CORFO (with support from the European Union [EU]) implemented a program in the Antofagasta region granting financial subsidies to large companies that participated in supplier development, requiring companies to take responsibility for the training and integration of suppliers (Program to Develop Suppliers for the Industrial Growth of Region II). CORFO then built on this program to support a collaborative effort across mines, involving ten large companies and two regional universities.

The Programa de Desarrollo de Proveedores (PDP) builds on these efforts by subsidizing a wide range of projects to increase supplier capacity: 50 percent of project costs are covered to the limit of US$10 million during the diagnostic stage (up to six months); and US$59 million with a cap of US$2.5 million per supplier (up to three years) during the development stage. Applicants must have sales of over 50,000 Unidad de Fomento (UF) per year (approximately US$2.4 million) and the supplier involved must have lower annual sales than the applicant and not exceeding 100,000 UF (approximately US$4.75 million). In addition to diverse programs such as PDP, the government of Chile also has more narrowly focused programs such as the Programa de Responsabilidad Social Empresarial Tributaria, which is explicitly about suppliers' adoption of electronic billing and business processing technology.

Financial Support
Financial support for suppliers can be provided in a number of ways to ease cash flow issues or to assist with capital growth. These include shortened payment terms; advance payment; and support for capital acquisitions in the form of providing contracts as surety, provision of capital, loans, or leasing schemes. Advance payment is one of the most common forms of financial support provided by the mining industry in the companies surveyed. In addition, some mining companies also offer financial support to local suppliers by prioritizing swift payment terms. This has the benefit of allowing the supplier to grow their business by alleviating working capital and cash flow problems. This is

a particular concern in the context of limited access to capital, which is a serious constraint for small businesses in Ghana and Mozambique. Newmont Ghana, for example, is committed to paying SMEs' invoices within 7 days, while the government of Chile has established a pro-SME award that is conferred on large companies that settle their payments with SME suppliers within 30 days—the ProPyme ("Pro-SME") Seal.[7]

Approaches to providing longer-term financial support include Anglo American's equity and loan investment program provided in South Africa (through Anglo Zimele) and Chile (through the Emerge Program). These programs, in addition to providing technical training and support, also provide an option for Anglo American to take an equity stake in the supplying company, should the business be deemed economically viable and strategically useful to the parent company.

Labor Market Effects

Evidence of Spillovers

Linkages

The direct employment contribution of mining to a host country is often small relative to the revenue contribution, given the capital intensity of most mining activities. The majority of spillovers occur when employees for foreign firms leave the company and use their know-how and skills to set up and/or work for local firms in the industry. However, in order for these benefits to be realized, local employees must be working in positions that give them access to knowledge, technological know-how, and managerial competences. Based on survey results, foreign firms in Chile consistently have much higher levels of local employees (figure 5.7). Mozambique, on average, is employing under half of Chile's level of local employees, which is likely to limit the potential for spillovers through labor turnover.

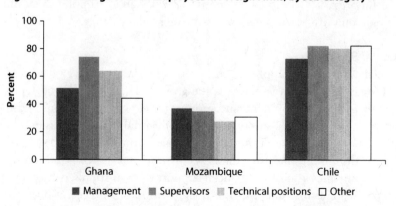

Figure 5.7 Percentage of Local Employees in Foreign Firms, by Job Category

Source: Surveys of 53 foreign firms and foreign suppliers: Ghana = 20, Mozambique = 16, Chile = 17.

The findings for Ghana and Mozambique are not consistent with the strict regulations in these countries limiting the number of expatriates permitted to be employed by foreign investors. This suggests that foreign-owned mining firms are negotiating with governments to grant further expatriate permits. These findings are reflected in the survey results, which indicate that in Mozambique (and to a slightly lesser degree in Ghana), a lack of suitable managers and supervisors and a difference in the level of expertise are perceived as significant obstacles to increasing the employment of local staff.

The potential for spillovers is enhanced through training and skills development, exposure to management, and technical practices and strategies, as well as the development of relationships and networks. In Ghana, foreign firms (mines and suppliers) indicated that a relatively high share of managers and supervisors have been promoted from within the organization—28 percent, compared to 18 percent in Mozambique and 12 percent in Chile, indicating a high level of skills and knowledge acquisition.

Actual FDI Spillovers through Labor Markets

Employees Joining Domestic Firms. Survey results showing previous employment for domestic suppliers indicate that in Ghana, over 60 percent of employees at domestic firms previously worked for other domestically owned firms across managerial, technical, and other positions. In Chile, over 80 percent of firms' staff also had worked at domestic firms. In contrast, Mozambican domestic firms receive a low proportion of their employees from domestic firms. This may reflect the relative vibrancy and depth of the industrial base in Chile, and to some extent in Ghana.

Across all the countries, Ghanaian domestic firms also have the highest share of managerial (15.8 percent) and technical (13.0 percent) employees who had previously worked at foreign firms—indicating some transfer of employees from foreign to domestic firms in the mining sector in the country. Data on the previous work experience of staff at Mozambican-owned suppliers also highlights that some employees at a managerial (13.1 percent) and technical (5.6 percent) level had some work experience at foreign firms.

In addition, the share of employees who previously worked for foreign firms in Ghana and Mozambique increases with the skill intensity of the job. This is evidence of these employees bringing knowledge that makes them capable of filling managerial and/or technical positions.

Employees Leaving to Establish Their Own Business. Positive spillovers can also be derived through labor turnover in cases where employees of foreign firms leave to start their own domestically owned businesses. In Ghana, a number of enterprises supplying the mining industry were established as a direct result of the entrepreneur having been employed within the mining sector. Not surprisingly, survey respondents indicated that only a small percentage of employees leave to start their own firms—in Ghana this was 1.7 percent of foreign firms (mining companies and suppliers) and in Chile 4.1 percent.

However, this small number can have significant benefits of broader job creation and capacity development.

In Ghana, key examples of firms established by former mine employees include:

- *Caesar Kilns and Furnaces*: A Kumasi-based company, manufacturing kilns and furnaces for the gold mining industry, which was established in 1997 by a qualified engineer with previous experience in a mining company. Based on many years of working in a mine's "gold room," he saw the opportunity to supply kiln products and maintenance services, which were previously imported and serviced by foreign suppliers. The company sold its first kiln in 2000, followed by the sale of some smaller kilns to small-scale mines, and has an ongoing role in maintenance and relining of the furnaces.

- *Electrofax*: A Takoradi-based electrical engineering company that provides design and supply of electrical infrastructure, mostly serving the mining sector. Established by an electrical engineer with 17 years prior experience with Ghana Manganese, the company's key projects have included electricity substation relocation and the installation of local electricity reticulation infrastructure. In 2007 he invested in a workshop to supply motor rewinding.

Interestingly, the survey results suggest a high proportion of employees at domestically owned firms in Ghana also left to start their own businesses: 15.8 percent reported in Ghana, significantly higher than the 1.7 percent reported in Mozambique.

Factors Contributing to Spillovers
GVCs and FDI Characteristics

Survey results indicate that the majority of foreign investors would like to increase the level of localization of management and technical staff. The main obstacle to doing this is perceived to be the lack of available local skills (see further discussion later in this section). Yet, differences in culture and management styles as well as concerns over protection of proprietary technology also stand out as being factors (figure 5.8); this is particularly the case in Chile.

The global skills gap in mining is driving mining companies to develop more extensive internal training programs. Some companies have also launched initiatives targeting the development of local skills.[8] For example, in Chile, Minera Escondida, through a nonprofit foundation, has initiated a specialized training center—the Centro de Entrenamiento Industrial y Minero—to support the development of mining occupational skills (related to, for example, electronics, electrical engineering, heavy machinery, and industrial machinery). Most survey respondents indicated that they provide training for employees (ranging from virtually all respondents in Ghana to two-thirds in Mozambique), with average days per year ranging from around 10 in Mozambique to 28 in Ghana (figure 5.9).

Figure 5.8 Importance of Obstacles to Increasing Employment of Local Staff

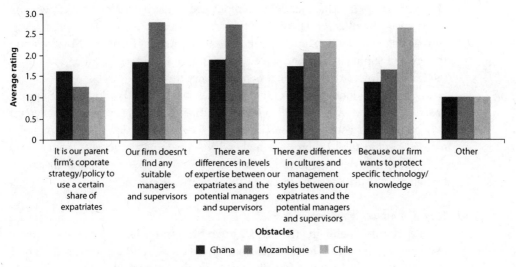

Source: Surveys of 14 foreign investors: Ghana = 4, Mozambique = 5, Chile =5.

Figure 5.9 Percentage of Foreign Firms Offering Training and Number of Days per Year

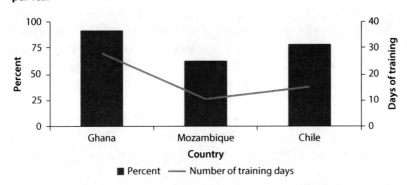

Source: Surveys of 55 foreign-owned mining companies and suppliers: Ghana = 20, Mozambique = 18, Chile = 17.

Where mining companies have a broad global base, they may promote exposure of employees to global operations in an attempt to provide standardized training across operations and ensure quality and consistency of employee knowledge. Exposure to an investor's global operations can also highlight company best practice for employees and develop leadership skills. One stakeholder explained that regional secondments within Africa were mandatory for staff identified as "high flyers" within the organization. AngloGold, for example, transferred its Ghanaian Managing Director of Iduapriem mining to Mali to run its two operations there in 2010.[9]

Domestic Firm Characteristics

Domestic firms in Ghana and Mozambique struggle to attract employees with experience gained at foreign mining firms, in particular where career progression and earnings expectations are lower. Furthermore, also in Ghana and Mozambique, the level of sophistication and scale of activities at a domestic firm appears to be significantly lower than a foreign-owned competitor.

In Chile, an innovation-focused project such as the World Class Supplier Program, which encourages collaborative solutions to mining operation challenges, may be more attractive for ex-employees of foreign firms, due to its cutting-edge nature. Also in Chile there is a higher level of domestic general managers of domestic firms (this was, in fact, even more pronounced for general managers of foreign-owned firms)—94 percent of firms, as mentioned previously.

Host Country Characteristics

Survey results highlight the very different approaches taken—particularly between Ghana and Chile—regarding requirements to hire local staff. Three quarters of respondents in Ghana indicated they were subject to requirements to hire local staff, while 40 percent of respondents did in Mozambique and none in Chile. Indeed, in Ghana, the Minerals Commission has developed regulations to promote inclusion of Ghanaians in mine workforces, including requirements for minimum proportions of Ghanaian staff in various employment and penalties for noncompliance. The Minerals Commission is working closely with mining companies to identify staff positions that can be filled by Ghanaians, preferred education and experience for these roles, local candidates that can be trained to replace foreign staff, required training programs, and targeted timings for Ghanaians to take up these positions. In 2009, 97 percent of 17,332 mine employees (increasing from 15,120 in 2000), were Ghanaian. Employment of Ghanaians at a senior level doubled from 1,505 to 3,180 between 2000 and 2011.

In Ghana and Chile, most respondents indicated labor turnover to be a problem. However, in all countries very few respondents indicated that labor turnover affects willingness to provide training.

Support for Spillovers

In **Ghana**, in addition to the regulatory approach, mining companies and foreign suppliers have made some efforts to support skills development. These include annual scholarships for training of mining engineers (via the Ghana Chamber of Mines), collaborations between the University of Cape Town and the Ghana Institute of Management and Public Administration to develop a new middle-management program (AngloGold Ashanti), and a six-month program to train Ghanaian engineers with the Tarkwa University of Mines (via Maxam). Despite these efforts, there remain significant shortcomings in education and training infrastructure and services in Ghana, which limit development of skills.

In **Mozambique**, the development of skills in the mining sector has been highlighted recently in the Human Resources Training Strategy for the Mineral Sector 2010–20 (Ministry of Mineral Resources 2010). The strategy aims to leverage private sector funding and public-private partnerships to supplement public resources to train 4,500 mining specialists by 2020. There were some initial collaborations in this area—notably between Vale and the government of Mozambique; however, significant skills gaps remain, particularly in educational and vocational training infrastructure and services.

Overall, **Chile** has a long track record of investing heavily in education and skills development and is seen by stakeholders as having reached an advanced level of skills and expertise. Currently, the Chilean government is playing an active role in addressing an expected shortage of 60,000 skilled workers in the mining sector, including skilled technicians, supervisors, and managers, as shown in box 5.10. Chile's universities are seen as having strong capacity, and the Universidad Católica del Norte in particular was highlighted as having strong capacity in mining education and skills development.

Demonstration, Competition, and Collaboration Effects

Evidence of Spillovers

Increased *competition* with foreign entrants in a number of markets (including product, labor, and credit markets) may result in upgrading by domestically owned firms in an attempt to keep up with foreign firms. For example, the Chilean mining supply chain has undergone significant expansion since the entrance of foreign mining companies. A number of domestically owned mining companies in Chile indicated an increase in both foreign-owned and domestically owned suppliers due to the presence of foreign investors in the mining sector. However, domestic mining companies in Chile, interviewed as part of this research, claimed that increased competition for resources competition actually had a negative impact.

Domestic firms may also upgrade through being *exposed to foreign firms' strategies and operations*, across sourcing, production, and marketing and distribution. In addition, FDI can present an opportunity to transfer proprietary technology and know-how around exploration and extraction that multinational corporations (MNCs) will only share with their affiliates through collaboration. Finally, *collaboration* on R&D can be across a wide range of areas. For example, in Chile, R&D collaboration ranges from across business management to engineering. Survey responses indicate higher levels of collaboration between foreign firms and public institutions in Ghana and Chile—while there is somewhat more limited collaboration in Mozambique, particularly with universities (figure 5.10).

The upgrading of HSE standards is one of the biggest areas of support given by foreign investors to their suppliers across all countries (see the first section of chapter 7, "The Apparel Global Value Chain"). However, demonstration effects can be observed more widely in supporting HSE upgrading in Chile. Stakeholders indicated that the emphasis placed on HSE by mining companies has led to a much broader awareness of, and application of, HSE standards, both within and outside of the mining sector. Stakeholders also see the entry of foreign mining companies as having created the need for HSE practices to be covered in mining

Box 5.10 Skills Developments for the Mining Industry in Chile

In Chile, the Ministry of Mines has collaborated on skills development with the Chilean National Service for Training and Employment—the Servicio Nacional de Capacitacion y Empleo (SENCE). A number of programs are in place to deal with the challenge of a skills shortage in the mining industry:

Training Program for Women in the Mining Industry
- Aims to train unskilled workers in specific mining-related areas, with a particular focus on equipment operation and maintenance.
- Each course provides 200–350 hours of training and is hosted by SENCE's technical training colleges.
- Over 4,000 applications were received for 1,330 training places in 2012.

Training for Work Program
- Program will provide 500 hours of training to 2,520 unskilled Chileans in Regions I–IV and helps them find work in the mining industry.
- Undertaken by SENCE's technical training colleges and offers courses in mechanical maintenance (42 percent of planned training spaces), truck and plant equipment operation (36 percent), and electrical maintenance skills (22 percent).
- Mining companies participated in program design to ensure the training is relevant for industry needs.

ENM (Mining Business School)—Universidad Católica del Norte
- Established in 2009.
- Joint effort between regional government, CORFO, mining firms, industry associations, and the university.
- Collaboration with Institute for Sustainable Mining at the University of Queensland.
- Offers a master's degree in mining management and a diploma in mine contract management.
- Modules delivered by key staff from mining companies.

Minera Ines de Collahuasi—Arturo Pratt University
- Constructed and donated a US$3 million Centre for Mining Technology.
- Provides an annual scholarship to the Centre for students in the region.

Codelco—Universidad de Chile
- Funding partner for the Department of Mining Engineering and for mining engineering PhD candidates.

Komatsu Cummings Chile—Universidad Tecnica Federico Santa Meria
- Signed a co-operation agreement in 2007.
- Aims to coordinate and develop education exchange activities including internships and training.

Figure 5.10 Foreign Investor R&D Collaboration

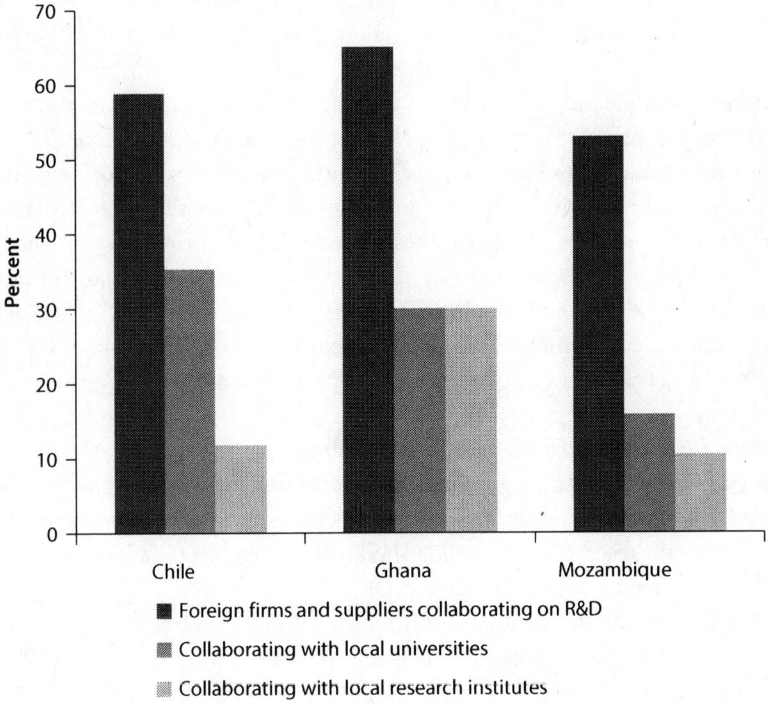

[Bar chart showing percentages for Chile, Ghana, and Mozambique across three categories:
- Chile: Foreign firms and suppliers collaborating on R&D ~59%, Collaborating with local universities ~35%, Collaborating with local research institutes ~12%
- Ghana: ~65%, ~30%, ~30%
- Mozambique: ~53%, ~16%, ~11%]

■ Foreign firms and suppliers collaborating on R&D
■ Collaborating with local universities
■ Collaborating with local research institutes

Source: Surveys of 56 foreign-owned mining companies and foreign-owned mining suppliers: Ghana=20, Mozambique=19, Chile=17.
Note: R&D = research and development.

legislation, which is seen to have led to much broader awareness and upgrading of HSE practice among local firms, in turn making them more competitive with foreign suppliers.

Factors Contributing to Spillovers
GVCs and FDI Characteristics
Emergence, and Upgrading, of Domestic Suppliers Due to Presence of Foreign Suppliers. Due to increasing levels of mining activity in certain regions, such as West Africa, as well as the movement toward higher-value service delivery by mining suppliers, there is a trend toward foreign suppliers setting up stronger in-market operations. This is particularly the case for equipment supply companies, where mining companies are increasingly contracting equipment suppliers on the basis of equipment performance (for example, tires).

Foreign mining services/contractors are also investing more in those markets that appear to have potential as a regional hub for the export of mining services into new markets, including Ghana and Chile. In Chile, for example, the Chilean offices of Fluor and SNC Lavalin have both developed new in-house capabilities to serve the region (Fernandez-Stark, Bamber, and Gereffi 2010), while SNC

Lavalin and Bechtel have also set up "Copper Centres of Excellence" in Santiago. These investments have been motivated in part by the potential for mining services in the wider South American market, particularly Peru and Colombia.

Upgrading of HSE Standards. Globally, the mining sector places a strong emphasis on maintaining high HSE standards. Due to hazardous working conditions, use of dangerous and toxic chemicals, and so forth, the mining sector faces intense scrutiny. Global "first tier" mining companies have a commitment to the "zero harm" principle, which commits mines to accepting no other performance level than zero and to full disclosure of their health and safety performance each year. The International Council of Mining and Metals, which has 22 of the world's leading mining companies as members, includes a continual improvement of health and safety performance as one of the ten principles of the organization.

Domestic Firm and Host Country Characteristics

Emergence, and Upgrading, of Domestic Suppliers Due to Presence of Foreign Suppliers. The primary drivers of sector linkages across the three countries are collaboration on research, developing the local supply base, and engaging in joint training and learning activities (table 5.6).

In **Ghana**, sector linkages among foreign mining companies are also driven by a need to lobby and engage with the government—particularly in the context of mining regulations on localization and local procurement. The Chamber of Mines supports some limited interaction between foreign and domestic suppliers through the Chamber's mining services group, which includes 43 mining suppliers—both foreign and domestically owned. In contrast, there is no institution in **Mozambique** that supports such interactions, although the "Mozambique Business Network" previously fulfilled a similar role. This is

Table 5.6 Drivers of Sector Linkages among Foreign Mining Companies

Factor	Overall ranking	Ghana	Mozambique	Chile
To collaborate in research	1	4	1	3
To develop local suppliers	2	3	5	1
To engage in joint training and learning activities	3=	2	5	5
To participate in joint corporate social responsibility or local economic development initiatives	3=	8	1	3
To approach potential customers together	5=	7	5	1
To share certain equipment/machines/infrastructure	5=	4	4	5
To lobby the government	7	1	5	8
We get better prices if inputs are bought/sourced jointly	8	4	5	8
We engage in subcontracting regularly or in peak times	9	8	5	5
To participate in joint product development	10	8	1	10
To participate in joint marketing activities	11	11	5	10
Other	12	12	5	12

Source: Surveys of 8 foreign-owned mining companies: Ghana = 3, Mozambique = 2, Chile = 3.
Note: Where a number is shown with an equal sign (e.g. "3="), this indicates that two or more types of support are ranked equally.

highlighted by the survey responses on the level of industry collaboration in Mozambique: only 40 percent of FDI mining firms reported having relationships with domestic firms in the sector, compared to 75 percent in Ghana and 60 percent in Chile.

In **Chile**, the institutional frameworks and relationships for supporting collaboration in the mining industry are seen as strong. A dense network of trade associations and supporting institutions encourage collective action and joint learning across the sector. Among these are the National Mining Council (Consejo Minero), representing the large mining companies and MNCs; Sociedad Nacional de Mineria, representing large and SME mining interests; and the Asociacion de Grandes Proveedores Industriales de la Mineria, representing the interests of large suppliers to the mining industry. In addition, regional industry associations in mining regions, such as the AIA or Corporación para el Desarrollo de la Región de Atacama, are key collaborators with regional government and national agencies such as CORFO. These associations are not affiliated solely with the mining industry and include wider stakeholders such as service providers, port operators, and local universities and training institutes.

Upgrading of HSE Standards. A focus on HSE standards in Chile is seen as a key driving factor for supporting upgrading of firms, as well as assisting firms to become globally competitive. Health and safety regulation is a key focus in Chile, where legislation such as the Supreme Decree 172 on mining safety regulations was approved in 1986 and updated in 2002 and 2004. Chile also has a strong network of nongovernmental organizations (NGOs) that promote improved HSE. This network includes the Chilean Safety Association (Asociacion Chilena de Seguridad, ACHS), a nonprofit organization established in 1957 to focus on risk prevention and minimizing occupational accidents across the economy; the NGO association Red Puentes (Bridges Network); and Accion RSE.

Support for Spillovers

There seem to be few efforts aimed specifically at supporting spillovers that may occur due to the presence of foreign companies through collaboration, competition, or demonstration in Ghana and Mozambique. In contrast, Chile promotes collaboration within the mining industry and value chain through a range of efforts.

In **Ghana**, the Chamber of Mines does, however, support sharing of information across its members. While its mining members are foreign owned, the Chamber's mining services group includes both foreign and domestically owned suppliers. The Chamber of Mines also assists Ghanaian entrepreneurs in the establishment of domestically owned mining operations, although it is unclear to what extent these efforts draw on experiences and operations of the foreign-owned mining operations. There are also some efforts under way to support upgrading of the artisanal and small-scale mining sector, for example, through supply of equipment to artisanal miners.

Chile has put in place a number of programs that support collaboration among companies, including on technology, research, and development. For example, the Mining Council set up a technology committee that defined high-priority areas where the industry aimed to stimulate research. This led to the establishment of a "Copper Technology Road Map" (AMIRA International 2004)—sponsored by domestic and international investors including Anglo American, Antofagasta Minerals, Codelco, BHP Billiton, and Rio Tinto—which defined R&D priorities for the industry. CORFO has also designed some of its support programs to encourage collaboration. For example, the Associative Development Project supports projects involving three or more companies that aim to increase their competitiveness in technical, production, financial, or managerial areas.

Conclusions

As shown throughout this chapter, it is clear that foreign investment in the mining sector—through supply chain effects, labor turnover, and competition, demonstration, and collaboration—has resulted in a range of spillovers for domestic firms. The subject is highly complex. It is rarely possible to demonstrate causality between the characteristics of a foreign firm, domestic firm, or host country and the level of spillovers to domestic firms when using descriptive statistics, surveys, and semistructured interviews. However, some important patterns emerge from the research.

The Level and Nature of Domestic Sourcing

Overall, the level of domestic sourcing is substantially lower in the Sub-Saharan African countries surveyed (Ghana and Mozambique) than in Chile. Among the main factors that appear to determine the level and nature of domestic sourcing in the mining sector are the following:

- Mineral commodities that tend to undergo a higher level of processing on-site, or more complex processes, present greater opportunities for local supply.
- Noncritical and less specialized inputs offer greater opportunities for local supply; however, spillovers may be limited due to the less sophisticated nature of these products and services.
- The trend toward centralized procurement may favor international companies; however, this may be offset by mining companies increasing their local procurement commitments.
- The structure of mining supply chains means that there is significant potential for local suppliers to engage in the supply chain as "secondary" suppliers, for example, as subcontractors to the primary contractor.
- More established, "top tier" or global mining companies are more likely than juniors to focus on linkages.
- Mining companies from source countries that speak the same language as host countries allow for much greater ease of communication and overall integration.

- The existence of more sophisticated domestically owned customers (mines or other) has typically helped domestically owned suppliers have a base level of sophistication, making it easier to meet the requirements of foreign-owned mining companies. With a larger gap in sophistication between domestic and foreign-owned firms, domestically owned suppliers would need to upgrade significantly to serve these customers—this increases the barriers to spillovers.
- Both the small size of individual domestic firms and the limited scale of total local production place constraints on increasing local sourcing and realizing growth opportunities in the supply.
- Inadequate quality control systems of some domestic firms limits further local sourcing and potential technical upgrading.
- Some of the concessions provided to mining companies as part of investment attraction packages may create disincentives to developing a local supply base.
- Countries with a strong legal framework and history of contract enforcement will encourage growth opportunities and further local sourcing.

Spillovers through Supply Chain Effects

Mine suppliers appear to benefit through technical upgrading and from growth induced by their relationship with foreign investors. While in Chile suppliers benefit through close relationships and strong, coordinated support programs, in Sub-Saharan Africa such support for suppliers is still emerging. Among the main factors that affect spillovers through supply chain effects are the following:

- Investors are more likely to support supply chain development where there is significant ongoing demand for the product. During mine development, there tends to be little emphasis on local procurement, whereas investors may support local supply chain development for products and services required for minerals extraction and processing.
- Companies with a local procurement policy, and with a clear track record of local procurement initiatives (and results) in other countries, are more likely to engage in supplier development.
- Mining companies from source countries that have a stronger focus on development tend to place greater emphasis on sourcing from companies with adequate HSE and quality standards, and therefore require (and in some cases support) upgrading of HSE and quality practices employed by local suppliers.
- Domestic firms can take advantage of potential spillovers not only through supply to mines, but also through relationships with their foreign suppliers based abroad (which in some cases may deliver greater benefits than supply with mining companies).
- Greater spillovers are realized when there is a more established mining industry in the country. Similarly, longer-lived mines offer a more sustainable market and give mining companies greater incentive to build capacity among local suppliers.
- Also over time, the development of mining supply hubs or clusters (for example, in Chile, Canada, and South Africa) fosters further development and increased

sophistication of products and services. This trend is supported by greater availability of skills, knowledge, and expertise as clusters develop.
- In Sub-Saharan Africa, clear policy and regulation to support local procurement and increased supply chain spillovers appears to have been important to facilitate the emergence of local suppliers. In contrast, Chile has achieved higher levels of local procurement than Africa with far less involvement of government regulation in the past 40 years, with government instead placing greater emphasis on supplier upgrading programs.
- Countries with strong intra-industry linkages and institutions—particularly those that include mining companies and suppliers—tend to have more local integration of mining supply chains due to higher levels of industry collaboration and joint product development.
- Having a strong mining association/Chamber of Mines can be critical in driving local procurement programs, including the identification of opportunities and associated local suppliers.
- Open investment and operating environments have a strong impact on the ability of local companies to benefit from mining supply chains.
- Countries with open financial markets where firms can access affordable finance are in a better position to invest for upgrading.
- Both governments and mining companies lack clear and consistently applied definitions of local procurement/local content.
- Support for improving access to opportunities is important for the development of supply links. However, technical support (in particular, support for improving product quality, product testing, and raising HSE standards) and financial support (in particular, improved payment terms, including up-front payment) are more likely to generate spillovers in terms of increased capacity of suppliers.
- Supply chain linkages programs (such as MozLink or the SPX program) have proved useful in establishing links between mines and suppliers. However, upgrading and competitiveness of supply is required for these relationships to be sustainable in the long term.
- Mining companies recognize the need to involve primary contractors in meeting local procurement targets and providing support to suppliers. However, there seems to be limited involvement to date of foreign-owned primary contractors in supporting domestically owned supplier development.

Spillovers through Skills and Knowledge Transfer

Skills and knowledge transfer through employment at foreign firms appears to be relatively high in Ghana, suggesting significant absorptive capacity and potential to benefit from spillovers. This is supported both by a base level of skills as well as targeted localization policy. Among the main factors that appear to affect spillovers through skills and knowledge transfer are the following:

- Firms that employ higher levels of local employees, particularly at senior levels, will increase the potential for labor spillovers as employees generate

knowledge and skills. Chilean mining companies employ, on average, 1.5 times the number of local employees at senior levels than foreign-owned companies in Ghana and twice that of foreign-owned companies in Mozambique.
- Foreign investors that have high levels of internal promotion and international exposure of local staff will also enhance their employees' skills and knowledge, which may support spillovers.
- Training focused on transferable skills, particularly financial literacy, leadership, and management, may support labor turnover spillovers as it can encourage entrepreneurship.
- Pre-operations training programs by foreign investors also increase the potential for labor turnover spillovers, as often they are established without any commitment by foreign investors to employ graduates of the programs.
- A lack of suitable managers and supervisors and a difference in the level of expertise are significant obstacles to increasing the employment of local staff, particular in Mozambique.
- Countries with high industry-university linkages will generate training courses that are more relevant and responsive to the needs of the mining industry.
- A history or presence of state participation in the mining sector may also lead to greater local skills acquisition.

Spillovers through Labor Markets and Entrepreneurship

The labor market appears to be a relatively limited source for spillovers in the case study countries. Among the factors affecting spillovers through labor markets and entrepreneurship are the following:

- Employment within the mining sector tends to be attractive relative to other sectors. Therefore, employees tend to stay at their companies rather than moving to employment in different sectors or starting up their own businesses. This inhibits labor turnover and associated spillovers. This may also have a negative impact on local skills availability in the rest of the economy, as mines attract and retain top local talent.
- Domestic firms in Africa often struggle to attract staff with experience working in foreign-owned mining companies. However, there is a realization among domestic firms that hiring such staff (and expatriates in particular) will lead to improved networks and links with foreign firms and development of globally competitive skills (including technical and management). These benefits in turn will lead to improved quality of delivery and business efficiency.
- Significant benefits can result from the establishment of new firms by previous employees of foreign mining companies. These new firms often become suppliers to the foreign investor that had previously employed the entrepreneur. This is currently only happening at very low levels, and entrepreneurs face many challenges in countries such as Ghana and Mozambique.
- In many Sub-Saharan African countries entrepreneurs face challenges in establishing businesses due to difficult operating environments and limited access to capital. This situation may constrain the number of employees leaving mining

companies to launch their own businesses. Furthermore, institutions supporting small business development are often weak and provide little support.
- There are isolated efforts around encouraging entrepreneurship within mining companies as a stepping stone to employees becoming contractors to the mining company.

Notes

1. Rupert Barnard, Michelle de Bruyn, Nick Kempson, and Philippa McLaren.
2. BRICS: Brazil, the Russian Federation, India, China, and South Africa.
3. The smelter has a capacity of 200,000 tons per year, but was idled in 2007 due to weak metal prices and electricity shortages.
4. The Fraser Institute's 2012–13 survey of mining and exploration companies confirms the importance of the current mineral potential (assuming current regulations and land use restrictions) for Chile and Ghana which plays a major role for mining companies in both countries compared to other countries (see Fraser Institute 2013).
5. http://www.transparency.org/research/cpi/overview.
6. http://www.programaemerge.cl/; www.angloamerican-chile.cl.
7. www.sellopropyme.gob.cl.
8. Sources include UNCTAD (2007), *World Investment Report*.
9. http://www.goldfields.co.za/sus_hr.php.

References

AMIRA International. 2004. *Copper Technology Roadmap*. AMIRA International. http://www.energetics.com/resourcecenter/products/roadmaps/Documents/Copper-Technology-Roadmap.pdf.

AngloGold Ashanti. 2012. *AngloGold Ashanti in Ghana*. http://aga.intervate.com/en/About-Us/Regionsandoperations/Ghana/Pages/default.aspx.

AreaMinera. 2010. "Drillco Tools: Alta Tecnologica nacional al servicio de la mineria y de la exploracion." *AreaMinera* 5 (46): 23.

Barnett, A., and M. Bell. 2011. "Is BHP Billiton's Cluster-Programme in Chile Relevant for Africa's Mining Industry?" Policy Practice Brief 7, The Policy Practice, Brighton, U.K. http://www.thepolicypractice.com/papers/18.pdf.

Barrick Gold Corporation. 2011. *Responsibility Report 2011*. http://www.barrick.com/files/responsibility-report/2011/Barrick-2011-Responsibility-Report.pdf.

Campbell, K. 2012. "Mining Stimulating Economic and Social Development in Mozambique." *MiningWeekly.com*, July 21. http://www.miningweekly.com/article/mining-and-miners-are-stimulating-economic-and-social-development-in-mozambique-2012-08-03.

Cochilco. 2006. "Desarrollo e Innovacion Tecnologica Minera en America." Organismo Ejecutor: Comisión Chilena del Cobre (COCHILCO) para IDRC – CRDI. Centro Internacional de Investigaciones para el Desarrollo. Informe Final, Abril de 2006 pp.149. http://www.tecnologiaslimpias.cl/chile/docs/11514356691Informe_COCHILCO.pdf.

Dona Ines de Collahuasi Mining Company. 2012. *Sustainability Report 2011*. Collahuasi, Chile. http://www.collahuasi.cl/pdf/sustentable/info_sustentable11_en.pdf.

Esterhuizen, I. 2011. "Concern over Proposed Amendments to Mozambique's Mining Law." *Mining Weekly*, December 9. http://www.miningweekly.com/article/concern-over-proposed-amendments-to-mozambiques-mining-law-2011-11-24.

Fernandez-Stark, K., P. Bamber, and G. Gereffi. 2010. *Engineering Services in the Americas*. Report commissioned by the Inter-American Development Bank & the Chilean Economic Development Agency (CORFO). Center on Globalization, Governance & Competitiveness, Duke University.

Fraser Institute. 2013. *2012/2013 Survey of Mining Companies*. Vancouver, Canada: The Fraser Institute.

Gold Fields. 2010. *Sustainability Report 2010*. http://www.goldfields.co.za/pdf/sustainability_reports/sustainability_report_2010.pdf.

Krause, M., M. Ackermann, C. Hirtbach, M. Koppa, L. Siciliano Brêtas, and L. Traub. 2008. *Business Development in Mozambique: What is the Role of the Regulatory Business Environment in Supporting Formalisation and Development of Micro, Small, and Medium Enterprises?* German Development Institute. http://www.die-gdi.de/CMS-Homepage/openwebcms3.nsf/%28ynDK_contentByKey%29/ANES-7KUHTK/$FILE/Draft%20Business%20Dev.Moz.pdf.

Liebherr. 2011. *Liebherr Annual Report 2011*. http://www.liebherr.com/Data/LHO/AnnualReport/2011/en-GB/files/assets/basic-html/toc.html.

Lomas Bayas. 2011. *Sustainability Report 2010*. http://www.xstratacopperperu.pe/ES/Publicaciones/Informe%20de%20Sostenibilidad/South%20America%20Operations%20Division,%20Peru%20-%202011%20Sustainability%20Report.pdf#page=3&zoom=auto,0,341.

Ministry of Mineral Resources. 2010. *Human Resources Training Strategy for the Mineral Resources Sector*. Government of Mozambique. http://www.mirem.gov.mz/relatorios/formacao/strategy_train.pdf.

Mlachila, M., and M. Takebe. 2011. "FDI from BRICs to LICs: Emerging Growth Driver?" IMF Working Paper WP/11/178, International Monetary Fund, Washington, DC.

Morris, M., and J. Fessehaie. 2012. "Global Value Chains and Supplier Development: Do Asian- and Northern-Owned Mines in Zambia Behave the Same?" PowerPoint presentation. PRISM, School of Economics, University of Cape Town. http://www.open.ac.uk/about/international-development/files/internationaldevelopment/file/Asia%20Symposoium%20Presentations/4_%20Mike%20Morris%2010_20%20-%2010_40.pdf.

Newmont Ghana. 2010. *Local Procurement Policy and Action Plan*. http://www.newmont.com/sites/default/files/newmont_ghana_local_procurement_policy.pdf.

Republic of Chile 1974. The Foreign Investment Statute. Decree Law 600. http://www.cochilco.cl/descargas/english/legislation/DecreeLa_n600.pdf?idNorma=1004876&buscar=Ley+20.363.

Rio Tinto. 2012. "Local Procurement Workshop." PowerPoint presentation delivered at the Local Procurement Workshop in Tete, Mozambique, August 21. Rio Tinto plc, London, U.K.

Sandvik. 2011. *Annual Report 2011*. Sandvik AB, Sandviken, Sweden. http://www.sandvik.com/Global/Investor%20relations/Annual%20reports/Annual%20Report%202011.pdf.

Teck Cominco. 2012. *Generations: Teck 2012 Annual Report*. Vancouver, Teck Resources Limited.

UNCTAD (United Nations Conference on Trade and Development). 2007. *World Investment Report*. Geneva: UNCTAD.

———. 2012. *Investment Policy*. Geneva: UNCTAD.

Urzúa, O. 2011. "World Class Suppliers for the Mining Industry." Presentation. BHP Billiton, Melbourne, Australia. http://www.commdev.org/files/2760_file_BHP_Escondida_IFC_Presentation_Washington_290611.pdf.

Vale SA. 2012a. *Sustainability Report 2011*. Vale SA, Rio de Janeiro, Brazil. http://www.vale.com/EN/aboutvale/sustainability/links/LinksDownloadsDocuments/Informacoes%20complementares_en.pdf.

———. 2012b. *Programa de Desenvolvimento de Fornecedores*. Vale SA, Rio de Janeiro, Brazil. http://www.abrhrj.com.br/site/Premio/2010/Vale_INOVE.pdf.

Warner, M. 2011. "Do Local Content Regulations Drive National Competitiveness or Create a Pathway to Protectionism?" Solutions Briefing #5, Local Content Solutions.

World Bank. 2012. *Increasing Local Procurement by the Mining Industry in West Africa: Road Test Version*. Report No. 66585-AFR. World Bank, Washington, DC. http://siteresources.worldbank.org/INTENERGY2/Resources/8411-West_Africa.pdf.

CHAPTER 6

Sector Case Study: Agribusiness

Kaiser Associates Economic Development Partners[1]

Abstract

Investment in the agribusiness global value chain offers a significant opportunity to raise productivity levels by adopting new knowledge, technology, and techniques, from farming through processing and manufacturing. Overall, the level supply chain, labor market, and other network linkages between foreign investors and the local economies are relatively higher in agriculture than in other value chains, driven by the fundamental requirement of sourcing domestic agricultural inputs. Yet significant differences exist across countries, particularly in the processing and manufacturing stages of the chain. Here, Sub-Saharan African countries appear to have less well-established linkages than in Vietnam, determined mainly by having less-sophisticated domestic firms as well as fewer commercial-scale farms. The increasing importance of global standards and certification appears to be a major catalyst for supporting knowledge transfer between foreign firms and domestic actors. Efforts to promote spillovers, particularly through input provision, financing, and technical support, are extensive throughout most countries. These include efforts that are government driven, foreign direct investment driven, and multi-stakeholder—many good sector-specific models appear to be available. However, their sustainability is less certain, underscoring the importance of complementary and cross-cutting policies to improve skills and address supply-side constraints to competitiveness in the sector.

The Agribusiness Global Value Chain

Introduction to Agricultural Global Value Chain

Agricultural value chains are highly product-specific, making it difficult to generalize across the sector. The wide range of commodities, their applications, and levels of processing and value-added affects the stakeholders, value chain dynamics, and potential for spillovers. In addition to traditional "food" chains, there is also a range of agricultural products, such as cotton, rubber, biofuel feed stocks, leather, and natural extracts, that enter into "nonfood" value chains. This can also affect the type of investment and potential spillovers in terms of the type of processing facilities, the extent of health and safety standards/ quality control, and the type of knowledge/technology spillovers. Given these caveats, three types of agricultural/agriprocessing value chains can be identified, each with specific characteristics: (a) primary food products, (b) processed food products, and (c) processed (nonconsumable) agricultural products. Table 6.1 outlines these three types of value chains.

As figure 6.1 illustrates, the agricultural value chain stages include cultivation and input supply, primary processing, secondary processing, and distribution and sales. Key activities at the cultivation stage of the value chain include input supply, cultivation, harvesting, and transportation. Actors in the cultivation stage include commercial farms, smallholders, cooperatives, and vertically integrated estates. The choice of cultivation model will depend on the commodity—some crops such as pineapples and avocados are better suited to plantation growing while others are more appropriate for smallholder production—and the decisions of the operating company.

In agricultural value chains, the extent of value added and complexity of agriprocessing activity can vary. Leaving aside informal and artisanal production, the main stages of activity include:

- *Primary processing*: Includes simple value-added activities such as grading, drying, cutting, and portioning products such as fresh fruit and meat products for primary consumable food. In this stage of the value chain there is limited product transformation. Actors include some commercial farms and farmer cooperatives, as well as multinational firms—and, in the case of the meat industry, state-owned export companies.

Table 6.1 Agribusiness Value Chain Types

Type of value chain	Examples	Characteristics
Primary agricultural products (consumable)	• Fresh fruit and vegetables • Staple food products (maize, rice) • Meat • Nuts • Tea	• One stage of "processing"—typically low technology processing (for example, packaging, cutting) • Limited product transformation before end-consumer
Processed agricultural products (consumable)	• Confectionary • Biscuits • Beverages • Dairy products • Coffee • Chocolate	• Two stages of processing—although can be vertically integrated • Substantial and complex product transformation • Branded/differentiated products
Processed agricultural products (nonconsumable)	• Textiles (cotton, leather) • Animal feed (maize) • Biofuels (feedstock) • Pharmaceuticals (palm oil, extracts) • Industrial use (rubber, various byproducts)	• Two stages of processing—although can be vertically integrated • Substantial product transformation • Often linked to waste products from consumable agricultural products (for example, biofuels, animal feed) • End-use in nonconsumable/nonfood products

Sector Case Study: Agribusiness

Figure 6.1 Overview of Agricultural Value Chains

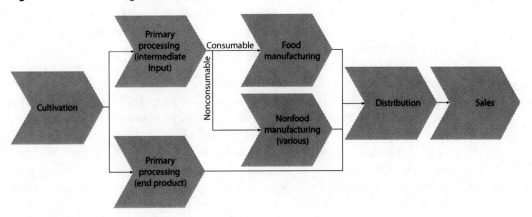

- *Secondary processing*: For processed food products, this involves food and beverage manufacturing activities such as bottling, blending, cooking, and mixing. In this stage of the value chain, there is significant concentration of actors, with multinational food and beverage manufacturers/branders leading the market. Other actors include local food and beverage manufacturers and contract manufacturers. Secondary processing can also include manufacturing activities for nonfood products entering other value chains, including biofuels, textiles, and pharmaceuticals, among others.

Lead firms in agricultural value chains vary depending on the commodity. In primary agricultural products such as fresh fruit, vegetables, and meat, the value chain is buyer driven. Retailers (particularly supermarkets) define what is to be produced, but focus on design, retail, and marketing activities rather than production (Gereffi 1994, 96). Global buyers have the ability to set strict process parameters and standards that producers must meet to engage in the value chain.

In food and beverage manufacturing, it is generally the brand owners, such as Nestlé, Kraft, Coca-Cola, and SABMiller that are the lead firms. The beverage industry in particular is highly concentrated. However, due to high concentration in the intermediate stage of the value chain, large commodity traders are also important actors in some product segments. For example, the cocoa value chain is dominated globally by three multinationals: Cargill, ADM, and Barry Callebaut. In other value chains (including fresh vegetables, coffee, and cocoa), smaller marketing agents may play an important role as "relational suppliers." They enable smallholder farmers to market produce by providing support with logistics, traceability and standards, supply of inputs, and some primary processing (for example, coffee milling).

Value Chain Dynamics

This section provides a brief overview of key dynamics in global agricultural value chains, with a view to understanding the impact that they have on the nature and level of foreign direct investment (FDI) spillovers. It looks at the trend toward standard in agricultural value chains, levels of vertical integration in the agricultural industry, and the types of governance found along the agricultural value chain.

Foreign Investment in Agricultural Land

Since 2007, there has been a significant rise in FDI in agricultural land in Africa and elsewhere in the developing world. Primarily, the investment is by resource-poor countries looking to secure food production for their own consumption, or to grow plantation crops such as oil palm and biofuel feedstock for large-scale industrial use. Investors in agricultural land include both private firms and, increasingly, governments, state-owned enterprises (SOEs), and sovereign wealth funds. This type of FDI is typically on long-term lease rather than land purchase and tends to focus on the production of basic commodities for export back to the FDI home country.

Significant debate exists around the rise in land-seeking FDI and on the benefits or disadvantages that may accrue to the host country and local communities. Serious concerns exist that such FDI may present risks to food security, undermine existing land claims, and displace local populations. At the same time, FDI investments in infrastructure and the introduction of modern farming practices have the potential to deliver substantial spillover benefits to local smallholders (many of whom may participate in the projects as contract farmers) and local communities.

Contract Farming

Globally, there is a trend toward sourcing agricultural products from smallholders through the use of contract farming and/or outgrower schemes, rather than pure plantation production, as agricultural investors adapt their sourcing strategies in Sub-Saharan Africa. The shift toward contract farming/outgrower schemes affects the potential for spillovers from FDI, as it brings a large number of suppliers into the supply chain, as well as benefits from technical assistance and financial assistance. The use of contract farming also increases the potential for spillovers from the foreign investor's perspective, as firms have a clear incentive to support improvements in terms of productivity, quality, and health and safety among their contract farmers.

Agricultural Standards

Global agricultural value chains are increasingly buyer driven. The lead firms can be seen as supermarkets, food manufacturers, or commodity traders, depending on the type of commodity. These lead firms exercise significant power over the value chain, but their control is characterized by passive forms of coordination. This is because complex product information has been embedded into widely accepted *standards and certification schemes* (Gibbon and Ponte 2005). These can be categorized as food safety/quality standards and social and environmental standards; typically they relate to products destined for mature export markets such as the European Union (EU), Japan, or the United States.

The application of a particular standard depends on the type of commodity and level of processing. Food quality and safety standards are the most common—although, increasingly, environmental, labor, and other ethical food standards are emerging in commodities such as coffee and cocoa, where UTZ[2] and Rainforest Alliance certification[3] is increasingly demanded by manufacturers in response to end-customer demands. Table 6.2 shows the range of standards that exist in modern food retail and their supply chains.

Social and environmental standards often also include modules on farming practices and food safety, but these standards are not exchangeable. Exporters that have Fair Trade certification may also have GlobalGAP certification. Food manufacturing operations in the agriprocessing sector are also required to have food safety standards such as hazard analysis and critical control points and good manufacturing practice. In addition to the importance of these standards for accessing export markets, and the implications they have for smallholder participation, there is evidence that these internationally required standards are also having an impact on supply chains in domestic markets. In Kenya, for example, there is some evidence that domestic supermarkets and wholesalers are beginning to require KenyaGAP (a localized variation of GlobalGAP) certification from their supply chains (Gereffi, Fernandez-Stark, and Psilos 2011).

The Rise of Modern Food Retail

Supermarkets have emerged as the key market channel for food retail sales since the 1990s. Research shows that the average share of supermarkets in food retail increased from 5 percent to 29 percent in 1990 to 50–60 percent by the mid-2000s in markets in South and Central America, East Asia (outside China and Japan) and Southeast Asia, Northern-Central Europe, and the Baltic countries (Reardon, Henson, and Bedegué 2007). There has been a trend toward consolidation within the supermarket segment since the 1990s, as large multinationals such as Wal-Mart, Carrefour, Tesco, and Royal Ahold have entered developing markets—particularly through joint ventures and acquisitions. In Sub-Saharan Africa supermarkets have spread rapidly over the past decade, with South

Table 6.2 Types of Agricultural Standards

Food safety	Product quality	Social/environmental
• Pesticide use and residue limits	• Grading	• Fair Trade standards
• Hygiene requirements	• Freshness	• Labor standards
• Traceability requirements	• Production composition	• Organic production requirements
• HACCP	• Product cleanliness	• Corporate social responsibility
• Food additives	• Labeling requirements	• Recycling requirements
	• Nutritional claims	• Animal welfare

Source: Gereffi and Lee 2009, 11.
Note: HACCP = hazard analysis and critical control points.

African chains (for example, Shoprite, Pick-n-Pay) leading the way, other retailers like Kenya's Nakumatt expanding across the continent, and multinationals recently entering (for example, Wal-Mart's acquisition of Massmart in South Africa in 2012).

The rise of supermarkets has also transformed supply chain governance. Supermarkets increasingly exert a high degree of control over their supply chains and set the conditions for other chain participants—including the adherence of suppliers to the required private standards. This has an impact on the participation of smallholders in the supply chain, and potential for spillovers. Evidence suggests that foreign retailers' procurement strategies in Sub-Saharan Africa can contribute to the exclusion of smallholders due to the costs of supply chain participation (including required investments and monitoring and certification costs) and economies of scale. On the other hand, standards also create a significant opportunity for upgrading of smallholders and other local firms. Meeting international standards should improve internal processes and contribute to improved competitiveness, and also provide immediate credentials that can lead to supplying other multinational corporations (MNCs) or to exporting.

Vertical Integration

There is some vertical integration between cultivation and processing in the agricultural value chain. Vertical integration is motivated by factors such as the need to secure access to commodities, customer-demanded traceability requirements, and attempts to enter perceived higher-value activities. Primary processors in several of the global value chains (GVCs) for traditional tropical export crops have also moved upstream and established purchasing companies in the countries of origin of the raw material. The degree of vertical integration depends on the agricultural product and the end-market. It can take place in a range of commodities including sugar processing, fresh produce, animal feed, and, more infrequently, food manufacturing. While vertical integration generally lowers spillover potential by internalizing production, in most cases in-house production is often combined with partial outsourcing of production and processing.

The Context for Spillovers: Commercial Production and FDI

Introduction

The countries under study have significant agricultural sectors across a diverse range of commodities. While agricultural production in all countries is still dominated by staple food crops, such as maize and cassava, FDI is directed at cash crops, typically for export. Table 6.3 shows key exports by value in these countries. Several products are common across most of them, including coffee, tea, sugar, cashew nuts, rubber, and palm oil. Also notable is the relative scale of Vietnam's agricultural sector—exports of Vietnam's 10 largest agricultural commodities is nearly three times the combined exports from the three African economies.

Table 6.3 Key Agricultural Exports by Value, 2010
US$, millions

	Ghana		Mozambique		Kenya		Vietnam	
	Commodity	Value (US$m)	Commodity	Value (US$m)	Commodity	Value (US$m)	Commodity	Value (US$m)
1	Cocoa beans	847.4	Tobacco	161.0	Tea	1,165.0	Rice	3,247.0
2	Cocoa butter	86.5	Raw sugar	107.3	Green coffee	187.9	Dry rubber	2,388.0
3	Palm oil	46.0	Pulses	28.2	Fresh vegetables	150.4	Green coffee	1,851.0
4	Sugar (refined)	41.0	Sesame seeds	26.9	Cigarettes	92.7	Cashew nuts	1,135.0
5	Cocoa powder	21.2	Cotton lint	26.7	Palm oil	83.6	Pepper	421.4
6	Beverages (alcoholic)	19.6	Wheat flour	17.2	Green beans	55.8	Fresh fruit	214.0
7	Natural rubber	13.6	Cashew nuts	17.0	Pineapples	55.1	Tea	200.0
8	Dry rubber	13.6	Bananas	11.3	Sugar (confection)	53.6	Dried cassava	196.3
9	Cashew nuts	13.3	Molasses	8.0	Beer of barley	35.3	Prepared food	110.1
10	Cocoa paste	11.8	Sugar (refined)	6.8	Preserved vegetables	33.2	Sugar (confection)	58.0
Total value		1,114.0		410.4		1,912.6		9,820.8

Source: FAOSTAT 2010.

Ghana

Overview

Agriculture remains an important contributor to the Ghanaian economy, accounting for approximately 40 percent of gross domestic product (GDP) and approximately 55 percent of national employment. The main agricultural products in Ghana are cocoa, horticulture, palm oil, and staple crops such as maize, yams, and cassava. There are also emerging developments in the cotton, cashew, and shea nut industries. The Ghanaian agricultural sector is dominated by smallholders with limited access to capital, technology, and markets. The cocoa industry in Ghana is by far the largest commercial agricultural subsector. Despite liberalization in 1992, the industry is still heavily regulated by the state-run Ghana Cocoa Board (COCOBOD),[4] which manages input supply, purchasing, and marketing of cocoa beans, and is a shareholder in the local processor, the largest licensed buyer, and a local fertilizer manufacturer. Horticulture, particularly fruits such as pineapples, mangoes, and bananas, are also an important subsector. The horticulture industry is dominated by large-scale plantation cultivation, and many of these plantations include outgrower schemes in addition to the nucleus estate. There is a significant foreign presence in the plantation industry, although there are some medium-size local exporters in the pineapple industry. In contrast to the fruit industry, the vegetable industry in Ghana is less developed, with limited exports—including chilies, Asian vegetables, and sweet potatoes—but significant potential for growth.

Foreign Direct Investment

In Ghana, FDI has been predominantly through fully-owned greenfield investments, although partnerships between domestic and foreign investors in Ghana and acquisitions of domestic firms have also been used as market entry strategies, according to survey respondents. Foreign companies have invested significantly in plantations across a range of products in Ghana, particularly in palm oil and horticulture (bananas and pineapples). Other FDI is involved in rubber and cotton production. Among planned investments, notable is the Kenyan vegetable producer VegPro, which is planning to invest US$3 million in a 1,000 hectare plantation to produce vegetables for export to Europe.

On the processing side, there is fairly extensive FDI in food processing (including Unilever, Nestlé, and Olam) and beverages (Heineken, Diageo, SAB Miller, Coca-Cola) with recent investments in palm oil refining, wheat milling, infant cereal, tomato processing, and biscuit manufacturing. Encouraged by a subsidy of 20 percent on the "light crop" bean price, all three of the global cocoa processing majors have invested in facilities in Ghana. Finally, Blue Skies (U.K.) processes cut fruit for the EU market and manufactures juice for the local Ghanaian market.

Across commodities, there is local participation at the cultivation stage of the agricultural value chain. Local participation takes place either through locally owned commercial farms, such as Bomarts, Koranco, and Milani in the pineapple industry, or through smallholder farmers, such as in the cocoa industry, which involves approximately 720,000 farmers in production. At the processing level in the cocoa industry, there are a number of local actors that have emerged in recent years—for example, Ghana Cocoa Processing Company and Plot Ghana—although most are struggling to remain profitable. In other agricultural areas, there is also local participation in palm oil processing industry (Appiah Menka) and dairy production (Emigoh, Emadom).

Kenya
Overview
In Kenya, the main agricultural products produced are tea, coffee, fresh vegetables, and cut flowers, as well as products for local consumption, such as milk, sugar, and maize. Foreign investors have focused on tea, coffee, horticulture, and floriculture.

- *Tea* is a very large foreign exchange earner, accounting for over US$1 billion in export earnings in 2010 and employing approximately 400,000 smallholders across the country (Export Processing Zones Authority 2005). Approximately 60 percent of Kenya's tea production is produced by smallholders and 40 percent by large tea plantations.
- *Coffee* is also an important agricultural sector, producing high-quality Arabica coffee, predominantly for export. Coffee is produced by both smallholders (55 percent) and coffee estates (45 percent).
- *Horticulture* includes both fresh vegetables and fruit, both of which are produced for export as well as domestic markets. Fresh vegetables are grown by a mix of commercial farms, outgrower schemes, and smallholders. Commercial farms also act as packers and exporters, and often have contracts

with supermarkets in the United Kingdom. These companies also have a number of outgrowers and contract farmers in addition to their fully managed plantations.[5] New small and medium enterprise (SME) exporters have also emerged and are supplying the European wholesale market. The fruit industry is dominated by pineapples, avocados, mangoes, and passion fruit. Foreign investors typically manage estates but there is a trend here too, toward expanding outgrower schemes.

- *Floriculture* exports began to grow during the 1980s, and are now one of Kenya's leading export sectors. The establishment of the flower industry was largely due to settler and foreign investment with strong ties to Western markets, with a number of firms run by Kenyans (including citizens of Indian origin) also participating in the industry.

Foreign Direct Investment

Many of the leading foreign agricultural supply companies are present in the market, including Bayer, Syngenta, and Yara. Plantation investments in Kenya include Del Monte's pineapple plantation near Thika and Kakuzi Ltd.'s (owned by U.K.-based Camellia Plc.) 14,500 hectare estate, which includes avocados, pineapples (under contract for Del Monte), macadamia nuts, seed maize, cattle, and timber. However, on the whole, resource-seeking investments in Kenya are less common due to the limited availability of land. There is significant foreign investment in the floriculture industry, with companies such as Oserian Farms (the Netherlands) and Finlays (the United Kingdom) investing in smaller, capital-intensive flower farms. There has also been substantial foreign investment in tea plantations, notably from Eastern Produce, Unilever, Williamson Tea, and Finlays. There is little foreign investment in coffee growing—only Kofinaf (a joint venture between Kenyan real estate developers and Renaissance Capital, a Russian investment bank) is active in the coffee plantation business.

Foreign investment in agriprocessing activities is limited to larger companies such as Nestlé and Unilever. Nestlé's factory in Nairobi is a small unit by global standards but manufactures brands such as Milo, Maggi, and Cerevita and also repacks coffee for the local market. Unilever's investments in agriprocessing facilities include tea processing factories and centralized manufacturing operations in Nairobi. Beverage manufacturers such as Coca-Cola and Guinness are also present with significant manufacturing/bottling operations. However, there is also evidence of multinationals exiting Kenya as production costs increase. Within the agriprocessing industry, a number of multinational firms including Cadbury have shut down their Kenyan operations and relocated to countries such as the Arab Republic of Egypt and South Africa in recent years (Nyabiage 2010).

The food retail industry in Kenya is dominated by local supermarket chains—namely Nakumatt, Uchumi, and Tusky's, and there is no foreign investment in the industry. Indeed, the Kenyan-owned supermarket chains are investing in the wider East Africa region. Kenya is notable for the high prevalence of modern food channels.

In all the main agricultural and agriprocessing sectors, there is strong participation of local firms producing both for domestic and export markets. Firms controlled by Kenyan citizens of Indian origin play an important role in the domestic agribusiness sector and maintain strong links to global markets.

Mozambique

Overview

In Mozambique, tobacco and raw sugar remain the main export crops, both of which are largely exported unprocessed. There are approximately 150,000 smallholder tobacco farmers in Mozambique with an average plot of 0.5 hectares (DRUM Commodities 2012). Only one tobacco-processing factory exists in the country. Production of sugarcane has increased rapidly since the 1990s, and currently covers more than 40,000 hectares in the country. While privatization remains an object, the government retains a small stake in each of the four commercial mills in Mozambique.

The cotton industry in Mozambique is regulated through a concession scheme where the government awards an area to prospective investors. These companies then have exclusive rights to purchase cotton from farmers in the region, and obligations to provide these farmers with inputs (seeds, insecticides) and technical assistance. Due to this system, cotton production is therefore also driven by approximately 300,000 smallholders, who account for 96 percent of cotton seed production (Chamuene, Mahalambe, and Catine 2010).

Other important agricultural commodities produced in Mozambique include new plantations of jatropha, and forestry, as well as vegetable production and processing for export. There is also high potential for fruit production, due to good climatic conditions, although at present, there is limited commercial fruit production.

Foreign Direct Investment

Foreign investment in plantations and agricultural estates is a relatively new development. Former government-owned sugar plantations were acquired by South African sugar manufacturers Illovo Sugar and Tongaat Hulett in the late 1990s. U.K.-based Aquifer has also invested in a 700 hectare plantation for vegetable production as part of the MozFoods/Vanduzi brand. Large-scale plantation investments in the forestry and biofuels sectors have also been made in the Nampula region—with land agreements granted to Norwegian-company Lurio Resources for a 126,000 hectare eucalyptus plantation and Italian firm Avia S.p.A. for a 10,000 hectare jatropha project, since 2008. However, the status of these projects is unclear at present. Portuguese firm Prio Agricultura (now Nutre Group) has also secured a 24,000 hectare concession for cereal and oil seed production in Sofala province—currently 3,200 hectares are under production. Olam International (Singapore) has also begun a rice farming initiative on 500 hectares in Zambezia province with plans to scale up to 10,000 hectares over five years. Other FDI in upstream agriculture in Mozambique includes Olam and Plexus in the cotton industry and Agriterra in the maize industry.

Sector Case Study: Agribusiness

FDI in agriprocessing and downstream agriculture is less developed. In the tobacco industry, Mozambique Leaf Tobacco is the only foreign investor. There are intermediate processing investments in the cotton industry by Olam International and Plexus (the United Kingdom). Similarly, South African sugar processors—Illovo Sugar and Tongaat Hulett—have both invested in upgrading of their sugar mills in recent years. Other agriprocessing investments in Mozambique include vegetable packaging operations for MozFoods (Aquifer—the United Kingdom) and investments in the maize supply chain by Agriterra Group (the United Kingdom). These include rural buying points, storage, and processing facilities.

FDI in Mozambique's food retail industry is dominated by South African supermarket chains.

Outside of smallholders, there is very little local participation in Mozambique's agricultural sector. Among the few large local firms are Citrum (citrus exporter), the Uniao Geral das Cooperativas (diversified cooperative including poultry), and Higest (poultry).

Vietnam
Overview
Rice is Vietnam's key agricultural crop for domestic consumption and export, and the country is the second largest exporter of rice by volume. Production has grown exponentially since the rice market was liberalized in 1989. Similarly, rubber production is also an important crop for export—with 85–90 percent of production exported in 2009. Production areas and yield for both crops have increased dramatically over the past decade. Another major export-earning agricultural subsector is coffee. Vietnam produced 1.1 million tons of low-quality robusta coffee in 2010, the majority of which is grown by smallholder farmers. Tea production is also important in Vietnam, with about 70 percent coming from smallholders and 30 percent from estates. The processing sector is expansive, including informal processors, private firms, FDI, and SOEs. Finally, there is an emerging cocoa industry in Vietnam; while still small it is being targeted for significant growth in the coming years.

Other commodities, such as fruit and maize, are produced for local consumption.

Foreign Direct Investment
The Vietnam market was only opened to foreign investment in the 1990s, and therefore most FDI has entered only in the last decade. In contrast to the Sub-Saharan African (SSA) country cases, where acquisitions were prevalent, most agribusiness investment in Vietnam has been greenfield.

FDI in the coffee exporting segment is common (Louis Dreyfus, Armajaro, and ED&F Man). As there are no foreign-owned coffee estates in Vietnam, FDI is limited to sourcing and primary processing. In terms of agriprocessing, there is substantial foreign investment from Western investors, such as Nestlé and Unilever, and from Asian investors from the Republic of Korea; Taiwan, China;

Japan; and Singapore. In particular, there has been significant foreign investment in the instant beverage segment. Foreign investors such as Olam, Nestlé, and Viz Branz have all invested in instant beverage facilities for soluble coffee production. There is also a strong foreign presence in the beverage manufacturing industry, with international players such as Heineken, SAB Miller, and Carlsberg conducting manufacturing in-country.

Local participation in the agricultural input supply segment of the agricultural value chain is relatively strong in the fertilizer industry and somewhat weaker in seeds. The food manufacturing segment in Vietnam is a mature market with a predominantly domestic market focus and there are numerous local enterprises across all products and a range of local enterprises across all agricultural sectors.

Drivers of FDI Location Decisions

Across the selected countries, key drivers of FDI indicated by survey respondents include political and social stability, an environment conducive to business, and access to raw materials (table 6.4). In Mozambique, investors also see the country as offering attractive fiscal incentives, while in Ghana access to land is a key driver of foreign investment. Additionally, investors in Kenya and Vietnam both see access to specific inputs as a key driver of investment—perhaps reflecting the higher level of industrial development in these two countries.

Furthermore, geographical proximity to raw materials and firm's customers were ranked as important factors in terms of geographical location of investment. The importance of proximity to customers in Ghana and Kenya may reflect the

Table 6.4 Ranking of Most Important Factors Driving FDI Location Decisions in the Agricultural and Agriprocessing Sector

Factor	Ghana	Kenya	Mozambique	Vietnam
Political/social stability	1	1	3	1
Business environment	2	5	7	5
Access to land/facilities	3	12	4	8
Access to reduced labor costs	4	6	7	3
Access to reduced nonlabor costs	5	1	9	10
Access to local/regional markets	6	9	2	5
Access to raw materials	7	1	4	4
Fiscal incentives	8	6	1	5
Access to specific inputs	9	1	9	2
Access to skills	10	9	11	12
Preferential market access	11	6	6	13
Personal relationships	12	14	14	13
Access to technology	12	11	11	15
Following our multinational customers	14	12	11	9
Other	15	16	16	16
Following our competitors	16	15	14	10

Source: Surveys of 29 foreign-owned agriculture and agriprocessing investors: Ghana = 11, Mozambique = 5, Kenya = 3, Vietnam = 10.

market-seeking foreign investment in these countries and the growing local market, in contrast to Mozambique, where this was not as important and where proximity to raw materials remains the key driver of investment.

Supply Chain Effects

Evidence of Spillovers

Linkages

The agribusiness sector in general tends to purchase more inputs from local markets than just about any other sector. However, there are clear differences across countries in the sample. Figure 6.2a shows that foreign firms in Vietnam reported sourcing more than half of their inputs from local producers (more than two-thirds, including locally based foreign producers). In contrast, less than 20 percent of inputs in Ghana and Kenya are reported to come from local suppliers, with much higher reliance on imported inputs. This likely reflects real differences in the availability and quality of the local supply base, although it is also partly a function of the mix of activities taking place in the country (or covered in the survey). Figure 6.2b highlights how the differences in local sourcing by foreign investors are shaped by the nature of the activity they undertake. As expected, the results shows that commodity traders' purchases from local firms are highest (67.4 percent), as these investors mainly purchase unprocessed agricultural goods. Interestingly, vertically integrated companies with some direct investment in agricultural plantations and some primary processing activities purchase the highest percentage of goods and services within the country (83.4 percent of total spending from both local firms and foreign suppliers in-country). This is because these investors often utilize outgrower/contract farming activities to supplement in-house production, and also have significant processing expenditures in terms of equipment, maintenance, and/or packaging materials. Food and beverage manufacturers purchase the lowest percentage of goods and services locally.

Of the goods and services sourced locally, raw materials, together with non-core services and packaging materials, account for the vast majority, although raw materials alone account for less than half the total (figure 6.3).

Actual FDI Spillovers through Supply Chains

Assistance by Foreign Investors. As part of supply chain relationships, foreign firms tend to provide farm and nonagricultural suppliers with support across a range of areas. Survey responses indicate that 100 percent of foreign-owned agricultural investors and 88 percent of foreign-owned suppliers provided some level of assistance to domestically owned suppliers. However, only 24 percent of domestically owned suppliers indicated that they received assistance from their foreign customers, and in Mozambique only one domestic supplier indicated that it had received any support at all. This may be explained by the fact that the domestic suppliers that responded were mostly nonagricultural suppliers, while foreign investors' main focus of support is for farmer groups. Some supplier

Figure 6.2 Sourcing of Goods and Services by Foreign-Owned Agriculture Firms (percent distribution by value)

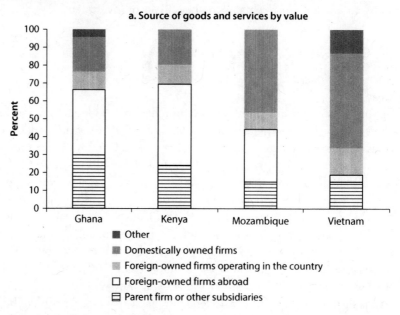

Source: Surveys of 74 foreign-owned agricultural and agriprocessing firms and agricultural suppliers: Ghana = 23, Mozambique = 8, Kenya = 22, Vietnam = 21.

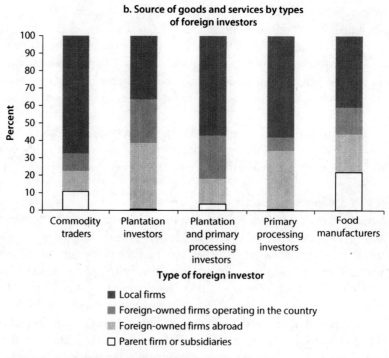

Source: Surveys of 30 foreign-owned firms across all countries: commodity traders = 7, plantation investors = 5, plantation/primary processing investors = 5, primary processing investors = 5, food manufacturers = 8.

Figure 6.3 Nature of Goods and Services Sourced Locally
percent

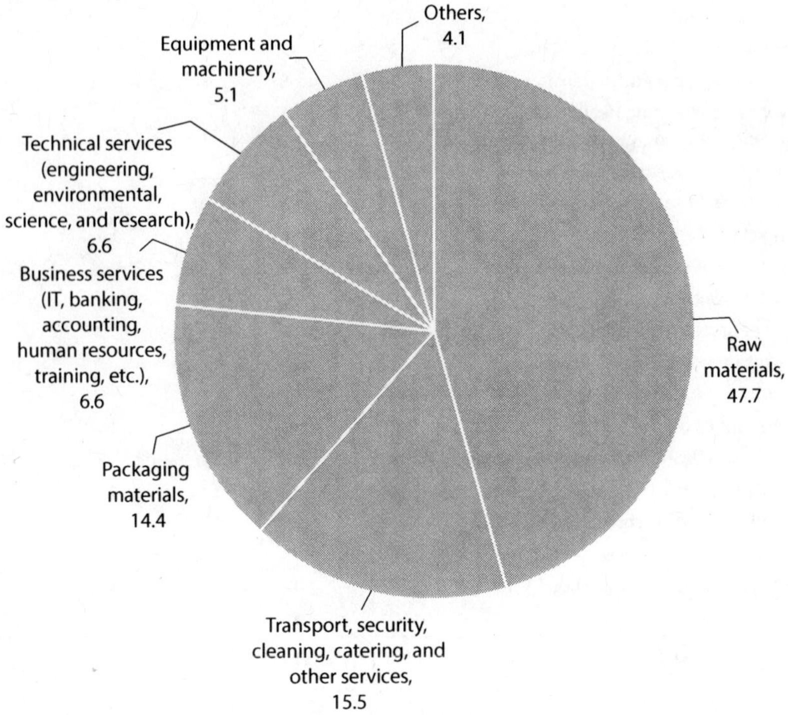

- Others, 4.1
- Equipment and machinery, 5.1
- Technical services (engineering, environmental, science, and research), 6.6
- Business services (IT, banking, accounting, human resources, training, etc.), 6.6
- Packaging materials, 14.4
- Transport, security, cleaning, catering, and other services, 15.5
- Raw materials, 47.7

Source: Surveys of 28 foreign-owned agricultural and agriprocessing firms: Ghana = 12, Mozambique = 3, Kenya = 3, Vietnam = 10. IT = information technology.

development for nonagricultural suppliers is enshrined in quality standards and certification, therefore mitigating the requirement for direct assistance.

Survey responses indicated that the most common types of assistance provided by foreign investors is provision of agricultural inputs and materials, advanced payment, and support around quality and health, safety, and environmental (HSE) issues (table 6.5). Access to funding and financing and worker training is also relatively common. Investors working with agricultural producers tend to focus more on financing and access to inputs, while those working with nonagricultural suppliers concentrate support for quality and skills. Interestingly, in Vietnam, foreign firms and domestically owned suppliers both rank support for "product or process technologies" assistance highly, suggesting that technology transfer may be more prevalent in Vietnam in comparison to the SSA countries under study.

Demand by Foreign Investors. Survey responses indicated that domestic suppliers in Mozambique (50 percent) and Kenya (36 percent) were requested to make significant improvements to their operations before or after signing contracts with foreign firms, while this was much less common in Vietnam and Ghana (19 percent and 10 percent respectively). Where improvements were

Table 6.5 Foreign Firms' Assistance to Suppliers, Ranked by Frequency of Support

Type of assistance	Overall rank	Ghana	Kenya	Mozambique	Vietnam
Provision of inputs	1	2	1	1	3
Advance payment	2	1	5	2	1
Support for sourcing raw materials	3	5	1	9	2
Help with implementing health, safety, environmental, and/or social conditions	4=	7	1	5	7
Help with quality assurance	4=	5	1	9	5
Support to get funds from other sources	6	2	9	5	8
Training of workers	7	4	5	9	11
Provision of financing for improvements	8	8	9	2	11
Help with inventory control	9	9	9	12	4
Product or process technologies	10	9	9	12	5
Repair/maintenance of machines	11	12	17	2	8
Help with finding export opportunities	12	17	5	5	13
Help with audits	13	12	5	14	17
Help with organization of production lines	14=	15	16	5	16
Financial planning	14=	12	13	14	13
Lending/leasing of machines or equipment	16=	15	13	14	13
Help with business strategy	16=	9	13	14	19
Licensing of patented technology	18	18	17	18	8
Other	19	19	19	18	18

Note: Where a number is shown with an equal sign (e.g., "4="), this indicates that two or more types of support are ranked equally.

required by foreign investors, improved quality control, increased volumes of production, and improved timeliness of delivery were the main improvements demanded (figure 6.4). Interestingly, while improved quality control was a key requirement of foreign customers, the survey responses suggest that foreign firms did not require this in the form of ISO 9000 or ISO 14000, possibly due to the existence of ISO 22000—a derivative of ISO 9000 that deals explicitly with food safety—or the extensive range of alternative quality standards that exist in the agriculture and food manufacturing industries.

Upgrading Technical and Business Capacity, HSE, and Quality. Through the types of assistance discussed, FDI can support farmers to operate at higher levels of productivity and align their production with market demands. Foreign investment in agriprocessing and food manufacturing also supports improvements in the technical capacity of domestic firms supplying inputs such as packaging materials, capital goods, and other services. The increasing complexity of certification schemes both for processed goods as well as for some raw materials (such as cocoa, coffee, and fresh produce) entering modern supply chains creates particularly good opportunities for local producers to benefit from interactions with foreign investors. Benefits include improved production techniques, the use of higher-quality inputs, and improvements in operational processes (for example, post-harvest handling, transport, and manufacturing).

Sector Case Study: Agribusiness

Figure 6.4 Improvements Requested by Foreign Customers before or after Signing Contracts: Percentage Requiring Improvements and Types of Improvement

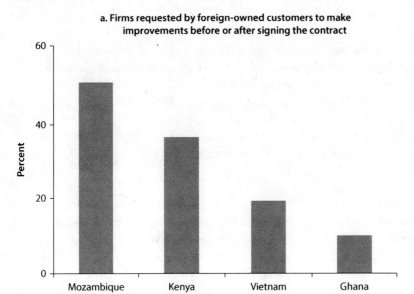

a. Firms requested by foreign-owned customers to make improvements before or after signing the contract

Source: Surveys of 68 domestically owned suppliers: Ghana = 10, Mozambique = 8, Kenya = 14, Vietnam = 36.

b. Types of improvements required by foreign customers (1 = seldom required, 4 = often required)

Source: Surveys of 16 domestically owned suppliers.

Examples of foreign investor support for nonagricultural supplier development in food and beverage manufacturing do exist. For example, in the Ghanaian cocoa processing industry, both Cargill and ADM indicated that they worked with packaging suppliers to improve their product quality (Cargill gave an example of supporting their pallet supplier to increase the durability of their pallets). However, explicit support appears to be less common than it is for farm-level supplier development, and support for suppliers of goods and services outside the areas of packaging and labeling also appear limited.

Upgrading through exposure to FDI is not limited to farmers and other input providers. In the food and beverage sector, distributors in the domestic market are also exposed to new technology used to support distribution systems, as well as good business practices for operating in modern distribution and retailing systems (see box 6.1).

Growth Opportunities. The majority of suppliers surveyed across the three countries indicated that they supplied more than one foreign-owned firm in the country—in many of these cases, they were referred by their initial foreign customer (figure 6.5a). Domestic suppliers in agribusiness have also been successful in leveraging FDI relationships to access export markets (figure 6.5b). This is

---

Box 6.1 Unilever's Support for Distribution Systems Upgrading

Unilever works through a multitier distribution and sales system, in order to maximize its market reach. For example, Unilever Ghana's distribution model involves 40 key distributors across the country, that then sell products to wholesalers/retail outlets. These distributors are formal businesses that employ, on average, 15 employees, and use modern technology systems. Unilever supports upgrading in distribution systems in two main ways:

- *Distributor training*: across a wide range of topics, including management and leadership skills, stock and inventory management, sales, and demand forecasting.
- *Introduction and provision of new technology for distribution:*
  - Globally, supplies hand-held terminals for stock management. These terminals are provided on credit to distributors and deducted from payments.
  - In Vietnam, installed SOLOMON software at distributors, which provides management support in a range of areas including product orders, sales, stock, and human resources.
  - In Kenya, bought licenses for store management software (shelf stocking, sales monitoring) for Kenyan retailers.
  - In Ghana, indirectly contributed to the adoption of new Global Information Systems technology through its outlet mapping service provider in Ghana.

Stakeholders in Ghana emphasize that a key spillover of these relationships is the "professionalization" of the businesses with which Unilever works. These softer skills can then be leveraged in other endeavors—for example, one Unilever distributor in Ghana has since diversified into hotel ownership.

Figure 6.5 Market Expansion by Domestic Firms Resulting from Supply to Foreign Customers

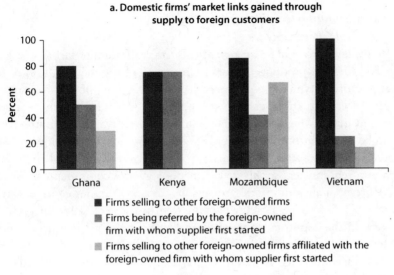

a. Domestic firms' market links gained through supply to foreign customers

- Firms selling to other foreign-owned firms
- Firms being referred by the foreign-owned firm with whom supplier first started
- Firms selling to other foreign-owned firms affiliated with the foreign-owned firm with whom supplier first started

Source: Surveys of 68 domestically owned suppliers: Ghana = 10, Kenya = 14, Mozambique = 8, Vietnam = 36.

b. Firms that became exporters as a consequence of supplying a foreign-owned firm

Source: Surveys of 44 domestically owned suppliers: Ghana = 9, Kenya = 14, Mozambique = 2, Vietnam = 19.

particularly prominent in the fresh produce segments. For example, in Kenya, the large volumes produced by foreign-owned investors have assisted in making direct air cargo links to Europe economically viable. As a result, domestically owned fresh vegetable companies are able to access export markets directly. Also, production by foreign floriculture operations of roses and carnations has allowed domestic producers, which tend to focus on summer flowers, to add their products to bouquets to supply to international markets. Other examples of successful export entry resulting from serving FDI in domestic markets include several firms in the research and consultancy sector (for example, Nagai Consulting in Ghana, and Allfield Services in Mozambique).

Factors Contributing to Spillovers
GVC and FDI Characteristics

The *level of outsourcing of production* affects the realization of farmer-level spillovers. For different products, factors such as ease of production, level of technology required, type of equipment/facilities required for processing, and product price fluctuations can affect whether products are suitable for smallholder and outgrowing schemes. For example, in Kenya, roses and carnations tend to be grown in-house by foreign investors, while summer flowers, which require less infrastructure and expertise, are grown by domestic firms. Coffee, which also requires less sophisticated technology and equipment, tends to be sourced from smaller, domestic growers. In turn, the type of contract farming undertaken will also be a factor contributing to the realization of farm-level spillovers. In the countries under study, centralized and nucleus estate models have been used extensively.

Stakeholders indicated a preference for medium- and large-scale outgrowers, where consistency is higher, transaction costs are less, and contracts are easier to enforce. This trend is likely to lead to greater collaboration between FDI and domestic firms, with higher potential for spillovers. On the other hand, these spillovers will be channeled to a narrower segment of the domestic industry. Evidence from the surveys indicates FDI in African agriculture still rely on a large bases of smallholders (the average firm in Ghana had 466 suppliers, in Mozambique 420, and in Kenya 373). By contrast, FDI in Vietnam appeared to be working with a much smaller set (44) or presumably larger-scale suppliers.

The increasing demand for *traceability and sustainability* in the supply chain of some agricultural commodities by lead firms and end-consumers is an important factor in realizing benefits in terms of market demand transmission to smallholders. Commodities where certification is required for supply chain participation or where certification can enhance the value of the commodity result in more participation by foreign investors to support improved productivity and quality of supply (see box 6.2).

The *motivation for foreign investment* and the *level of in-country processing* also affect the extent to which supply chain spillovers are realized. Market-seeking foreign investment is the only type of FDI that will result in the realization of technological upgrading by distributor networks in-country; market-seeking foreign investors are also more likely to engage in collaborative technical development with suppliers and subcontractors. The degree to which resource and efficiency seeking investment in agriculture deliver supply chain spillovers depends on several factors, including: (a) whether investors use *outgrower or plantation models* (the latter are more likely to lead to spillovers); (b) the *target markets* to which they sell (those targeting markets with demanding quality requirements—for example, the EU—are more likely to assist input providers); and (c) the degree of *vertical integration* of foreign investors (more vertically

Box 6.2 Cocoa and Tea: Certification Examples

UTZ
- UTZ-certified cocoa increasingly is a requirement of global chocolate manufacturers.
- Mars Inc. has a commitment to use sustainably grown cocoa in 100 percent of products by 2020.
- Nestlé has committed to using only source-sustainable cocoa for KitKat production by 2014.

Rainforest Alliance
- Unilever aims to procure all Lipton tea sold globally from sustainable sources by 2015.
- Nestlé's "Nescafe Plan" launched in 2011 commits Nestlé to purchasing 90,000 metric tons of Rainforest Alliance–certified coffee by 2020.

---

integrated investors rely less on domestic suppliers, and hence have less incentive to develop their capacity).

Company size and strategy may also determine the extent to which supply chain spillovers are realized. Large multinational firms are increasingly attempting to define themselves as partners within their supply chain, in order to present a positive corporate image globally. Two prominent examples are Olam and Nestlé, whose mottos are "Creating Value is our Business" and "Creating Shared Value" respectively. These firms also have substantial resources, capacity, and experience through which to develop and deliver supplier development programs.

Finally, the *nature of supply relationships* between domestically owned suppliers and foreign customers has an important impact on the realization of supply chain spillovers. Survey responses indicate that only 24 percent of domestically owned suppliers received assistance from their foreign customers. However, of these, suppliers that entered into formal contracts with their customers were more likely to receive support than suppliers who had only ad hoc relationships with their customers. This suggests that better established, long-term supply relationships will create a greater opportunity for the realization of technical upgrading spillovers, as assistance to suppliers becomes less risky and exposure to foreign requirements and business practices increases (table 6.6).

Domestic Supplier Characteristics

Survey responses from foreign investors suggest that poor delivery and low levels of technology are the key obstacles to foreign firms increasing their local sourcing, as shown in table 6.7. In addition, in Ghana and Kenya, insufficient production scale is identified as the biggest obstacle. Interestingly, quality and HSE issues are not seen as significant barriers to establishing supply relationships. This suggests that training and other assistance can relatively quickly address any existing gaps in those areas, while the obstacles that truly affect supply relationships are more fundamental and would require much more intensive support.

Table 6.6 Types of Suppliers Receiving Assistance from Foreign Customers

Type of relationship	Number of suppliers	Receiving assistance from foreign customers (%)
Ad hoc relationships	25	16
Contracts	44	25

Source: Surveys of 69 domestically owned suppliers.

Table 6.7 Types of Obstacles to Further Local Sourcing by Country
1 = biggest obstacle

Factor	Overall rank	Ghana	Mozambique	Kenya	Vietnam
Unable to deliver on time	1	2	2	1	5
Low level of technology	2	5	1	3	3
Lack of trained personnel	3=	7	2	3	1
Unable to finance necessary investments	3=	3	5	3	2
Production capacity too small	5	1	5	1	8
Pricing is uncompetitive	6	3	5	3	5
Inadequate quality control	7=	6	10	8	3
Poor engineering and design capacity	7=	7	5	8	7
Does not have required quality certification	7=	10	5	3	9
Cannot meet HSE standards	10	9	4	8	9
Other	11	11	11	8	9

Source: Surveys of 30 foreign-owned agricultural and agriprocessing firms: Ghana = 12, Mozambique = 5, Kenya = 3, Vietnam = 10.
Note: Where a number is shown with an equal sign (e.g., "7="), this indicates that two or more types of support are ranked equally.

The often *informal nature of domestic firms*, particularly at farm level, also affects the realization of spillover benefits due to the difficulties in enforcing contracts. Informality means that foreign investors have less security, but that they will be able to source products from farmers to which they have provided assistance. For example, the inability to enforce contracts with smallholder farmers in Kenya allows farmers to side-sell their produce (that is, sell produce to firms other than the firm that provided inputs and credit). In Indonesia, Armajaro stopped the prefinancing of cocoa farmers precisely because of the high level of side-selling to local traders (VECO 2011).

There are, however, some cases of strong links to spillover benefits, often where there is upfront payment or provision of seedlings by the investor—particularly Olam's operations in Ghana and Mozambique. Indications from investors are that the provision of upfront support to farmers in terms of chemicals and seedlings seems to be much greater in African countries, perhaps due to lower levels of commercial awareness compared to farmers in Vietnam.

Host Country Characteristics

Availability of farmland and land tenure systems influence the extent of supply chain spillovers at a farm level, because it determines the initial investment decision and the nature of investments (for example, plantations/estates versus sourcing from smallholders and commercial farmers). In all the countries being studied, land ownership by FDI is prohibited and business entities are limited to long-term leases of agricultural land for a period of 50–99 years.

Access to (and availability of) finance is one of the most significant characteristics of a host country and affects the ability of both farmers and manufacturing enterprises to take advantage of supply chain spillovers in the agricultural value chain. The costs associated with improving operations (either though the purchase of higher-quality inputs, certification, or improving manufacturing practices) can be high. This is particularly so for rural farmers, who often do not have formal bank accounts, collateral, or a credit record, and to whom commercial banks may be unwilling to grant loans. Similarly, formal manufacturing enterprises also struggle to access affordable financing to upgrade their processes and facilities to meet the standards or scale of operations required to engage in agricultural supply chains led by foreign-owned firms. With interest rates at 20 percent or more (as in several of the case study countries), it is difficult for suppliers to service loans, particularly with output that is being sold on export markets, where suppliers may not benefit from domestic inflation.

Availability of certification providers is also an important host country characteristic to support certification at a farm level. A local certification body can reduce the costs and time scales associated with certification. In Kenya, there are two accredited certification bodies for GlobalGAP—AfriCert and the Kenya Flower Council (KFC). In contrast, there are no certification bodies in Ghana or Mozambique. The benefits of local certification bodies are demonstrated by the fact that approximately 2,569 farms in Kenya are GlobalGAP certified, compare to only 244 smallholders in Ghana.[6]

Support for Spillovers
Enhancing Access to Supply Chain Opportunities

Programs that support supply chain spillovers by enhancing domestic farmers' access to opportunities include government regulation to ensure smallholder participation, and programs that support directly the establishment of local supply chain linkages. In Indonesia, regulation has been put in place, using a "nucleus-plasma" model, specifically to ensure participation by smallholder farmers in crop production, and upgrading of smallholder farmers through links with large—mainly FDI—palm oil estates (box 6.3).

Other programs used to support accessibility to supply chain opportunities have included business linkages programs (for example, through the United Nations Conference on Trade and Development) and similar development (for example, the West African Sorghum Chain Development). Outside of programs to support smallholders, however, there have been few efforts at facilitating supply linkages in other parts of the agribusiness value chain.

Technical and Financial Support

Public sector support is also an important factor in the realization of farm-level spillovers in terms of certification and improved productivity. In Ghana, Kenya,

Box 6.3 Palm Oil: Indonesia's Nucleus-Plasma Model

Smallholder participation in the palm oil industry in Indonesia has been driven by the government of Indonesia's cooperative program, Perkebunan Inti Rakyat (PIR-Trans), which encourages large-scale plantation owners to support smallholder development. Under the scheme, the "nucleus" plantation company can access low-cost loans from the government to develop their core (or *inti*) farms, and in return are expected to plant a surrounding "plasma" that can then be divided into smallholdings (2 hectares for palm nut and 1+ hectares for other crops) and allocated to local farmers. The nucleus plantation company assists the plasma farmers to develop and manage their smallholdings until ownership can be transferred to the smallholders. Farmers are obliged to sell produce to the *inti* once their plantations are mature.

Overall management stays with the nucleus estate until setup costs[a] and interest are recovered through subtracting 30 percent of smallholder processing returns. Upon repayment, plasma smallholders receive the land titles and a cooperative takes over service provision. The government of Indonesia determines conversion costs and debt owed by smallholders. The period of repayment varies from 5 to 14 years due to factors such as soil fertility and farming capacity.

The government's support for plasma farming linkages has also seen two variations of the PIR-Trans plasma model outlined above:

- *Koperasi Kredit Primer Anggota* (Members' Primary Credit Cooperative), applied to the palm oil sector after 1995. This model works with formal cooperatives rather than individual farmers. Financing is provided to the cooperatives for plasma development, with the *inti* company providing a corporate guarantee.
- *Pola Patungan* (Joint Venture model). This model was introduced as a way to stop conflicts over variable productivity of blocks and to support greater efficiency. This model gives smallholders share certificates for 2 hectares, rather than for a specific block of land. Shareholders have two options: (a) work in the plasma under the cooperative and be trained by the plantation company, or (b) work in the nucleus staff.

The nucleus-plasma model has seen 900,000 hectares of smallholder palm oil established, with improved incomes for smallholders in mature nucleus-plasma schemes. Another outcome has been the development of independent palm oil smallholdings, which can produce at a lower cost relative to estates.

The system has, however, faced a number of challenges, including: limited capacity of farmers, continuing low levels of technological expertise, low productivity, cooperative mismanagement, delays in receiving land and loan financing, difficulties around decision making with the plantation company, and restrictions on traditional intercropping.

Sources: Zen, Lubis, and Edyono 2005; Vermeulen and Goad 2006, 55.
a. Recoverable costs include: loan expenses, harvesting, transportation, fertilizer.

and Vietnam, the public sector, through research institutions and agricultural extension services, has played an important role in the establishment and upgrading of production activities. Various public sector organizations (including commodity-specific organizations such as COCOBOD in Ghana) provide support in the form of: analysis of suitability for crop production, varietal testing, and provision of seedlings and other inputs to farmers. In the coffee industry in Vietnam, the public-private partnership Coffee Task Force plans to use agricultural extension officers to scale up the use of demonstration plots for sustainable coffee production.

Given the large number of smallholder farmers in Ghana, Kenya, and Vietnam, training and certification programs must be implemented at a significant scale in order to make a meaningful impact. Foreign investors have taken a number of approaches to scale up delivery of technical assistance and training to farmers (see box 6.4). These include:

- Use of demonstration plots to support uptake of good farming practices, for example Unilever's plans to establish demonstration farms in Vietnam's tea sector
- Establishment of farmer training centers
- Formation of groups and "train-the-trainer" approaches
- Use of media and local languages to reach a wider target audience

---

Box 6.4 Coffee and Cocoa: Technical Support Programs

Atlantic Commodities—Farmer Training Centre (Vietnam)
- Establishment of a farmer training center and demonstration plot for coffee production on a four hectare site
- Curriculum (developed by Atlantic Commodities) comprises 10 modules including fertilizer application, shading, composting, and health, safety, and environmental issues
- Deliver training on a weekly basis, to a total of about 3,000–4,000 farmers
- Established by Atlantic Commodities (a subsidiary of ECOM Agro-Industrial Corporation) and supported by an International Finance Corporation loan
- Collaboration with fertilizer companies to deliver chemicals-related training

Cargill—Farm Field Schools Program (Côte d'Ivoire)
- Operates 300 programs across Côte d'Ivoire, with a capacity to train 25,000 cocoa farmers annually
- 10-month program implemented in partnership with Solidaridad (Dutch NGO) and ANADER (Côte d'Ivoire extension service)
- Curriculum covers a range of topics including fertilizer use, harvesting, pruning, post-harvest practices, disease prevention, and farm renewal

In addition to agronomic assistance, some foreign investors offer support for *improving business practices and governance* to foster improved management competency at the farmer level (see box 6.5). Providing this kind of training can support the realization of improved productivity and quality by supporting farmers to effectively manage the delivery of their produce and improve productivity through better organization and governance. Mechanisms include direct training and financial incentives to support collective selling. These programs are typically delivered alongside technical support to improve agricultural practices, rather than as separate programs, and also include support on how to reduce wastage, which is a major issue in developing country agriculture.

Methods of implementation of technical support also vary, but most programs involve *partnerships with various role players*. These partners can be nongovernmental organizations (NGOs), government ministries or public extension services, food manufacturers/branders, or a combination of these. Some foreign investors have also established their own NGOs and special-purpose vehicles to provide technical assistance to domestic farmers (box 6.6).

As noted earlier, advance payment and the provision of inputs are the most common forms of assistance provided by the foreign investors in the companies surveyed. This is most likely due to a limited access to finance and a limited cash flow for rural farmers, who are often unable to finance inputs such as seeds and chemicals at the start of the growing season. A number of foreign investors offer *prefinancing of inputs* such as seeds and chemicals, often on an interest-free basis. This is particularly critical for initial investment in tree crops such as coffee and fruit, which have a long production cycle, and for concession-based crops such as cotton, where there is an incentive for the investor to promote higher yields.

Alternatively, some foreign investors provide corporate guarantees that allow local farmers to access bank credit (instead of prefinancing). For example, Ghana Rubber Estates Ltd. supports outgrowers to access loans from the Agricultural Development Bank to start up their rubber plantations as part of their Rubber Outgrowers Plantation Project, while Armajaro provides guarantees to farmers in Indonesia (after the risk of side-selling meant that pre-financing was not viable) (VECO 2011; Paglietti and Sabrie 2012). Foreign-owned customers also support

---

Box 6.5 Coffee: Governance and Organizational Capacity Building

Armajaro—VECO (Indonesia)
- Armajaro offers financial incentives to a local farmer organization in Indonesia to promote collective selling. The farmer organization receives an additional fee of US$10 per metric ton if quality standards are met, which it uses to purchase fertilizer and inputs for members.

Technoserve—Cooperative Governance Support (Kenya)
- Provides support and training to 40 cooperatives on governance issues, farm-level bookkeeping, and sustainability.

Box 6.6 Special-Purpose Vehicles for Technical Assistance: The Source Trust (Armajaro)

- A not-for-profit organization set up by Armajaro to leverage traceability standards to strengthen the connection between multinational companies and farmer communities.
- Enabled by traceability systems, the premiums paid by these companies are used to support projects in the communities that produce the cocoa.
- Partners include multinational chocolate companies, such as Lindt, Ferrero, Meiji, Toms, and Camile Bloch, and donor organizations to support delivery of technical assistance to farmers.
- Technical support aims to improve quality, crop yields, and business knowledge of farmers.
- Projects funded by the "source trust" include specific agricultural activities such as farmer training/extension services; seed nurseries to provide high-yielding hybrid seeds; and farmer shops to provide inputs such as chemicals, machinery, and farm clothing.

Wider community projects, such as Village Resource Centres, malaria prevention programs, and the provision of community infrastructure are also supported.

---

domestic suppliers through *improved payment terms*. In Vietnam, Armajaro offers immediate payment for coffee sourced from domestic coffee buyers, often on a "free on truck" basis. This allows buyers to lower working capital and turn-around stock much faster, along with strengthening their relationships with farmers.

Labor Market Effects

Evidence of Spillovers
Linkages
Sector experience, whether gathered through employment with foreign companies in-country, or through business-to-business interactions with foreign firms, allows individuals to gain skills, knowledge, and business networks, which can then be applied to starting up new businesses or utilized by ex-employees who are re-employed by domestic firms. The degree to which such spillovers are possible is predicated on the local staff being employed in positions that will give them exposure to this knowledge.

Survey results in figure 6.6 show that local staff account for the majority of management and technical positions across all countries, suggesting relatively good prospects for spillovers through labor market channels. Localization of management ranges from a low of 62 percent in Ghana to a high of 78 percent in Vietnam. Overall, Mozambique shows the lowest levels of localization, with local staff accounting for 62 percent of supervisory staff and 59 percent of technical staff; Kenya and Vietnam have the highest levels of localization, with an average above 80 percent across the three key categories of staff.

Figure 6.6 Percentage of Local Employees in Foreign Firms, by Job Category

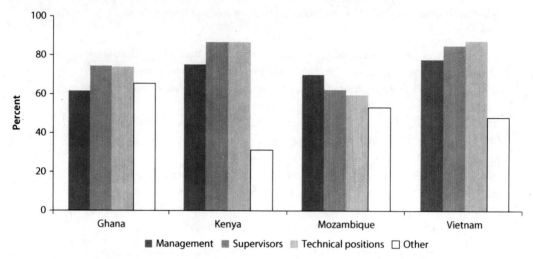

Source: Surveys of 72 foreign firms and foreign suppliers: Ghana = 22, Kenya = 22, Mozambique = 7, Vietnam = 21.

Actual FDI Spillovers through Labor Markets

Employees Joining Domestic Firms. Table 6.8 shows that domestically owned investors in Ghana (15 percent) and Kenya (10 percent) have the most employees with experience in foreign-owned firms, while in Vietnam and Mozambique the average percentage of current staff with previous experience in foreign firms is only 1 percent and 2 percent respectively. The corresponding figure for domestically owned suppliers is higher across all countries but Kenya. Twenty percent of managerial and technical employees in Ghanaian-owned suppliers are reported to have had previous work experience at foreign firms. This indicates that there is a significant circulation of labor across foreign and domestic firms, creating opportunities for spillovers.

However, semi-structured interviews and secondary research confirmed a limited evidence base for labor turnover spillovers from foreign investors to domestic firms overall, as foreign investors remain attractive employers within developing companies due to high salaries, job security, and potential for career progression.

Employees Leaving to Establish Their Own Business. Positive spillovers can also be derived through labor turnover in cases where employees of foreign firms leave to start their own domestically owned businesses. Survey responses show that former employees of foreign companies that leave to start their own businesses are not particularly common, although some evidence suggests significant entrepreneurialism in Ghana. Some anecdotal examples are presented in box 6.7.

Sector Case Study: Agribusiness

Table 6.8 Average Percentage of Staff with Foreign Firm Experience, by Type of Domestic Firm

Type of domestic firm	Ghana	Kenya	Mozambique	Vietnam
Domestically owned investors	15	10	1	2
Domestically owned suppliers	20	8	8	12

Source: Surveys of 82 domestically owned suppliers and 42 domestically owned investors.

Box 6.7 Examples of Entrepreneurship Linked to FDI in the Agribusiness Value Chain

Eden Tree (Ghana)
- Direct rival fresh produce exporter set up by previous employee of Blue Skies

Josra/Joswam/Ransley Coffee (Kenya)
- Three Kenyan-owned coffee exporters
- Started by former employees of international coffee traders such as Ibero and Dormans

Kandia Fresh Produce (Kenya)
- Set up by two previous employees at Everest (not a foreign company, but shaped by knowledge of foreign buyers and networks established through working in the export sector)
- Sells into EU wholesale market rather than to supermarkets

Factors Contributing to Spillovers
GVCs and FDI Characteristics

Firm strategy toward the use of local managers may vary by country of operation, based on, for example, the nature of FDI, language issues, and availability of skills locally. Survey results showed that across all countries, the majority of foreign investors would like to increase the percentage of local managers employed by their firm (which currently ranges from 60 percent to 80 percent across countries). However, obstacles to local employment in Vietnam and Mozambique, particularly the low level of expertise, limited availability of managers, and differences in culture and management style, are seen as major obstacles to increasing local employment (figure 6.7). In Ghana however, all these issues are only minor obstacles to increasing local employment.

The *quality and extent of training* offered by FDI affects the potential for realizing spillovers through labor markets. The majority of survey respondents indicated that they provide training for employees (figure 6.8a). In addition, labor turnover was not considered to be a significant problem facing foreign investors, and does not appear to affect firms' willingness to invest in training (figure 6.8b). Foreign investors in Ghana noted that losing staff to other foreign investors was common, particularly when new foreign entrants enter the market. However, this was considered standard practice and did not constrain spending on training.

Figure 6.7 Importance of Obstacles to Increasing Employment of Local Staff

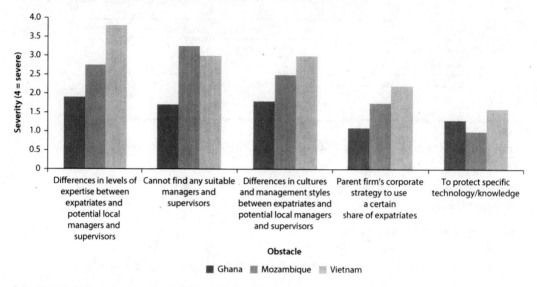

Source: Surveys of 19 foreign investors: Ghana = 10, Mozambique = 4, Vietnam = 5.
Note: Obstacles are rated on a scale of 1–4, where 1 means "no obstacle" and 4 means "severe obstacle." Kenya is excluded from the figure due to a lack of data.

Domestic Firm Characteristics

The more sophisticated are domestic firms, the more likely their employees will be in a position to access and absorb more advanced technologies and knowledge. Therefore, business sophistication as a whole is the most critical firm-level determinant of realizing spillovers through all channels. The level of training offered by domestic firms can also play a role. Figure 6.9 indicates that domestic suppliers in Kenya and Vietnam are more likely to offer staff training than in Mozambique and Ghana, although firms that do offer training in Mozambique report the highest intensity in terms of days of training provided.

Host Country Characteristics

Strong university-industry linkages support labor turnover spillovers through the establishment of industry-driven, relevant skills development. For example, foreign investors in the agriculture and agriprocessing industry have been involved in the establishment of university course design in Ghana and Kenya (box 6.8). These programs aim to respond to specific skills gaps in the sector through private sector input in curriculum development and industry placements. The two examples that follow have been established without any commitment by foreign investors to employ graduates of the programs. Therefore, there is potential for these skills development programs to benefit local firms or encourage graduates to establish their own businesses.

More widely, the existence of relevant training institutes will also support skills development in the food manufacturing and agriprocessing segments.

Figure 6.8 Quality and Extent of Training Offered by Foreign Firms, and Effect of Labor Turnover

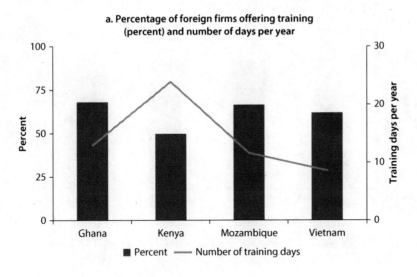

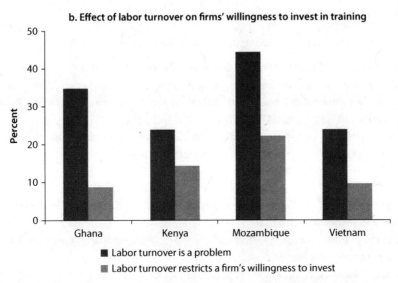

Source: Surveys of 74 foreign investors and suppliers: Ghana = 22, Kenya = 22, Mozambique = 9, Vietnam = 21.

Survey responses suggest that relevant public or private training facilities do exist and are utilized in Ghana and Kenya, where over 75 percent of foreign firms indicate that they use existing public and private training facilities (figure 6.10). In contrast, less than half of respondents in Vietnam considered national training facilities to be relevant. In Mozambique the utilization rate is reported to be low despite high relevance.

Figure 6.9 Percentage of Domestic Firms Offering Training and Number of Training Days per Year

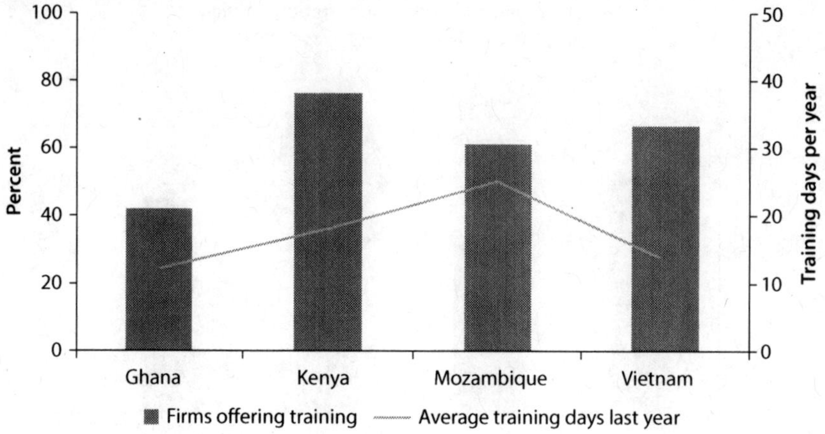

Source: Surveys of 88 domestically owned suppliers: Ghana = 19, Kenya = 17, Mozambique = 13, Vietnam = 39.

Box 6.8 Industry-Driven Training Programs in Kenya and Ghana

Coffee Diploma—Kimathi University College of Technology (Kenya)
- Industry-wide collaboration between the Coffee Board of Kenya, Kenya Coffee Producers and Traders Association, the Kenya Coffee Trader Association, the Coffee Research Foundation, and the Kimathi University College of Technology
- Two stages: Stage 1: certificate in coffee technology and quality management (6 months); Stage 2: diploma in coffee technology and cupping (12 months)
- Involves 3–6 month placement within the industry

Agricultural Logistics Skills Development Program—Kwame Nkrumah University of Science and Technology (Ghana)
- Funded by Archer Daniels Midland and Safmarine (international shipping company) in conjunction with the World Cocoa Foundation
- Training in logistics, transportation, supply chain, warehousing, and information technology
- Qualifies participants as logistics and warehousing technicians

---

Support for Spillovers

There are limited specific programs supporting spillovers through labor turnover across all the countries analyzed. Support programs generally focus on wider objectives, such as supporting establishment and growth of SMEs, skills development, or increasing the transferability of existing agricultural and agriprocessing skills. Public programs supporting SME development and wider skills development—particularly those funded by donors—appear to be more

Figure 6.10 Utilization of Existing and Relevant Training Facilities by Foreign Investors and Foreign Suppliers
percent

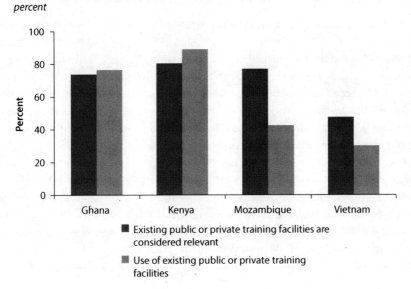

Source: Surveys of 74 foreign-owned firms and foreign-owned suppliers: Ghana = 23, Kenya = 21, Mozambique = 9, Vietnam = 21.

prevalent than privately funded programs (although see box 6.8 for information on industry-led training programs in the cocoa and coffee industries in Ghana and Kenya).

In terms of skills development, support programs or projects are also supported by donor funding—either through direct programs such as CBI's Export Coaching Program or through collaboration projects such as the Horticultural Practical Training Centre in Kenya (box 6.9). The Centre will support skills development across the horticultural industry by providing training to domestic firms and cooperatives, suppliers to foreign investors, and investors' internal staff.

Support for skills development in agriculture can also be delivered through the retroactive certification of existing skills in the industry. This kind of support can enable transfer of employees between companies by providing a common certification for all employees that is recognized across the industry and can improve the relevancy of public training programs. One example of this is the labor skills certification system initiated in the horticulture industry in Chile in the 1990s (see box 6.10).

Demonstration, Competition, and Collaboration Effects

Evidence of Spillovers

Foreign investors in production and marketing can actively promote *demonstration effects* to support upgrading of their suppliers through the establishment of demonstration plots or nucleus farms, where farmers and outgrowers can see firsthand the benefits of adopting better farming practices. Stakeholders note that

Box 6.9 Fresh Produce: Horticultural Practical Training Centre, Kenya

The Horticultural Practical Training Centre was established through a public-private collaboration between the Fresh Produce Exporters Association (FPEAK), the Kenya Agricultural Research Institute (KARI), and the Horticultural Crops Development Authority. The construction of the Centre was funded by a grant from the Dutch Government. The Centre was set up as a response to the problem of ad hoc capacity building and aims to centralized horticultural training. The Centre provides training to large-scale commercial farmers, smallholders, and cooperatives across a range of horticultural products, and practical demonstration on all horticultural-related subjects from land preparation to post harvest handling and value addition.

A separate grant was awarded to a Dutch university for a joint capacity-building project between the Centre, Jomo Kenyatta University of Agricultural Technology (JKUAT), and KARI to strengthen organizational capacity, develop and implement educational programs, and develop research capacity. The project will review curricula for agricultural training at the Centre and JKUAT and develop training materials. Industry stakeholders will have input into JKUAT curriculum through FPEAK in order to improve the relevancy of the industry.

Box 6.10 National Labor Skills Certification System for Horticulture, Chile

Established in 1988, the National Labor Skills Certification System was set up to provide a framework for recognition of skills across 15 industries—including horticulture—in conjunction with private investors. The program documents the skills and competencies required for jobs in horticulture production, packaging, cold storage, and processing, and provides a system of certification for existing workers. For example, a packaging employee can receive a certificate outlining their packaging skills, or in the processing industry, workers who operate machinery will receive accreditation of their ability.

Certification facilitates the transferability of skills in the horticultural industry (particularly among temporary workers), decreases uncertainty in the hiring process, and accredits workers' abilities. To date, 9,000 workers have been certified.

demonstration is particularly effective at the farm level, as farmers tend to be risk averse and suspicious of new methods and practices. Similarly, foreign input suppliers—whose customers are both foreign investors with commercial farms and outgrowers/suppliers to foreign investors—also actively demonstrate the benefits of using their products through demonstration plots (see box 6.11). In addition to demonstration from foreign investment in-country, domestic firms also learn through demonstration effects from foreign investors outside the country.

Figure 6.11 shows that the demonstration effects such as learning about new technologies and machinery, learning about marketing strategy, and learning about production processes are important spillovers for domestically owned firms competing in the same segment as foreign investors. In addition, a number

Sector Case Study: Agribusiness

Box 6.11 Demonstration Effects—Formal and Informal

Demonstration Plots: Coffee Task Force (Vietnam)

The Coffee Task Force is a public-private partnership between the government of Vietnam and foreign investors such as Yara and Nestlé. The program establishes pilot demonstration plots promoting sustainable coffee production and has resulted in yield increases, improved coffee bean quality, and an increase in farmer profitability. Due to its success, there is now a proposal to scale up 2 existing demonstration plots to 50 demonstration plots across three provinces. In-kind loans will be provided to government extension officers in the regions to establish demonstration plots.

International Demonstration Effects: Bomarts and Del Monte (Ghana)

Bomarts—a pineapple exporter—plans to upgrade its cultivation techniques after observing the productivity levels, technological advances, and equipment utilization of foreign investors in pineapple plantations. These improvements were identified after an international research trip to Del Monte's Costa Rican plantations. This trip was organized by HPW AG (a German pineapple distributor and marketer with a strong relationship with Del Monte in Europe) but funded by Bomarts. Stakeholders indicated that the research trip allowed Bomarts to learn extensively about pineapple cultivation and processing, and they intend to apply these lessons in their Ghanaian operations.

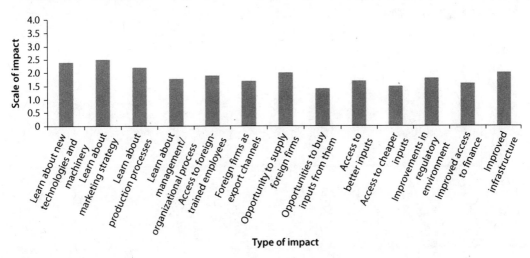

Figure 6.11 Perceived Positive Impacts of Foreign Investment on Domestically Owned Firms
1 = no impact, 4 = strong impact

Source: Surveys of 32 domestically owned firms: Ghana = 9, Kenya = 8, Mozambique = 3, Vietnam = 12.
Note: Impacts are rated on a scale of 1–4, where 1 means "no impact" and 4 means "large impact". Domestic firms surveyed may not be active in the same segments and/or commodities as the foreign firm responses.

of foreign agriprocessing firms indicated that the presence of their products in the marketplace has led domestic firms to develop their products, and in particular their packaging, in order to compete with foreign investors' products.

Entry of foreign investors can result in increased *competition* between local and foreign firms in a number of areas—most importantly in product markets, where foreign and domestic firms compete for sales in the same segment. Foreign investment may also result in increased competition between domestic firms and foreign investors in the labor and credit markets.

Figure 6.12 shows that in the agricultural sector, domestic firms across all countries perceive that the strongest impact of foreign investment has been to increase competition in the industry. Domestic firm responses also indicated that foreign firm entry has also had some impact on the industry wage level and competition for land and natural resources.

Across countries, there are some differences in the impact of foreign investment on local firms. In particular, domestic firms in Vietnam, Ghana, and Kenya perceived that foreign investment in the country had led to a loss of market share and a loss of customers. In Mozambique—where foreign investment in the agricultural and agriprocessing industry is at a more nascent stage—domestic firms consider this impact to be less significant. Domestic firms in Mozambique and Vietnam in particular consider that foreign investment has made it more difficult

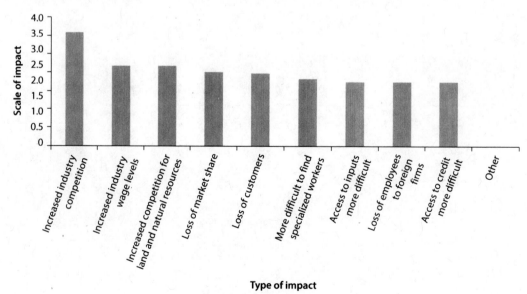

Figure 6.12 Perceived Negative Impacts of Foreign Investors on Domestically Owned Firms across Countries
1 = no impact, 4 = strong impact

Source: Surveys of 32 domestically owned firms: Ghana = 8, Kenya = 8, Mozambique = 4, Vietnam = 12.
Note: Impacts are rated on a scale of 1–4, where 1 means "no impact" and 4 means "large impact". Domestic firms surveyed may not be active in the same segments and/or commodities as the foreign firms, affecting responses.

to find specialized workers, while Ghanaian and Kenyan firms did not consider this to be one of the more significant issues.

Collaboration between foreign investors and domestic actors—particularly at the horizontal level—can also result in the technical upgrading of domestic firms as knowledge is transferred between companies. Collaboration appears to be common in the agricultural input segment of the value chain, particularly in the seed industry. Collaboration is also common in the food manufacturing industry, where investors' business models include contract manufacturing. In agricultural production, there is no less evidence of collaboration between competing foreign and domestic firms outside of industry associations.

Evidence from the surveys suggests that collaboration is relatively common across the African sample countries, while it is much lower in Vietnam. However, the collaboration that is taking place in Vietnam bridges across foreign and domestic firms, whereas foreign investors in Africa are mostly interacting with one another. Reasons for collaboration also differ across countries. Collaboration in Ghana is motivated by the desire to share equipment, machines, and infrastructure. In Kenya, participation in a joint corporate social responsibility initiative was an important reason for collaboration. In Vietnam, the key reason for sector linkages was engagement in subcontracting.

Survey responses suggest that there is some level of collaboration between foreign investors and public institutions on research and development activities across all countries. Respondents noted that research and development (R&D) collaboration covered areas including technology development, farming practices, laboratory analysis, and skills development.

Factors Contributing to Spillovers
GVCs and FDI Characteristics
FDI investment motivations are one of the most important factors determining the degree of investors' strategic interest in collaboration. Table 6.4 shows clearly that both access to technology and access to skills are among the least important determinants of investment across all four countries. This suggests there is unlikely to be much strategic interest for collaboration, particularly on research, among foreign investors.

GVC characteristics, including the *level of industry concentration and scale of foreign firm operations*, will affect how far domestic firms can realize technical and strategic improvements from observing foreign firms' activities or reacting to competitive pressures. In value chains where there is a concentration of large global firms at the processing level (for example, coffee and cocoa) some demonstration effects that are observed by domestic firms cannot be realized. For example, one local cocoa processor in Ghana has attempted to improve operations at a marketing level by cultivating direct relationships with large multinational customers in response to the activities of large multinationals such as Cargill and ADM. However, given the GVC dynamics of the cocoa industry, in which multinational customers require cocoa products from a range of countries for flavor stability and require large volumes, the local firm must sell through a cocoa broker, thus eroding the firm's profitability. In this case, the Ghanaian processor has been able to identify the process improvements as demonstrated by the foreign investors, but has been unable to implement the required changes.

Firm strategy in food manufacturing also influences how far improvements through collaboration can be realized. In cases where contract manufacturing is observed in the food manufacturing segment, the potential for spillovers is greater. Contract manufacturing requires foreign investors to share technical specifications and proprietary information with local firms and also requires domestic manufacturers to meet foreign investor requirements in terms of cost, quality standards, and health and safety.

The *R&D strategy and location of R&D activities* for foreign investors is an important factor affecting the realization of productivity increases deriving from collaboration between foreign firms and local research partners, particularly in the international seed industry. Foreign investors tend to conduct the majority of their R&D activities in centralized, global research centers (often in their home countries). Some R&D is undertaken in strategically located regional research facilities, and, to a smaller extent, at a local level. MNCs such as Pioneer, Monsanto, Pannar Seeds (South Africa), and SeedCo (Zimbabwe) all have small research stations in Kenya, but their main research centers and seed production units are located outside the country.

Domestic Firm Characteristics

The *size of domestic firms* will affect how extensive technical improvements can be in response to demonstration and competition effects. Larger domestic firms will be able to realize economies of scale similar to those of foreign investors, and will therefore be price competitive with these foreign firms and have lower costs relative to smaller firms.

The *technology and business sophistication* of domestic firms is also a key factor in the realization of technical upgrading in response to demonstration and competition effects in the agriprocessing and food manufacturing sector, as well as collaboration on joint product development and R&D. Domestic firms with high levels of technological competency and business sophistication do not need to make such large changes to their business model in order to compete with foreign firms effectively, which will mean that spillovers are more easily realized but may be smaller in scale. Figure 6.13 shows that domestic firms' perception of the technology gap are highest in Ghana, while in Vietnam the gap between domestic and foreign firms is perceived to be less.

Host Country Characteristics

Intra-industry relationships and linkages are an important factor for the realization of spillovers from demonstration, competition, and collaboration effects. Institutions and forums for interaction will increase the visibility of demonstration effects and improve social capital between actors for collaboration. Figure 6.14 suggests that relationships between firms in the agricultural and agriprocessing

Figure 6.13 Perceived Technology Gap between Domestic Firms and the Top Foreign-Owned Competitor
1 = no gap, 4 = large gap

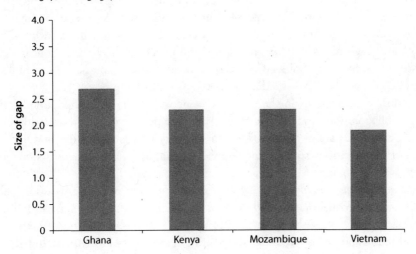

Source: Surveys of 43 domestic producers: Ghana = 10, Kenya = 12, Mozambique = 4, Vietnam = 17.
Note: Perceived technology gaps are rated on a scale of 1–4, where 1 means "no gap" and 4 means "large gap." Domestic firms surveyed may not be active in the same segments and/or commodities as the foreign firm responses.

Figure 6.14 Percentage of Firms That Have Relationships with Other Firms Producing Similar Products

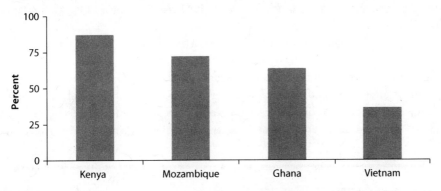

Source: Surveys of 76 domestic producers: Ghana = 22, Kenya = 16, Mozambique = 11, Vietnam = 27.

sector are higher in the African economies under study than in Vietnam, where intra-industry relationships are considerably lower.

The strong intra-industry linkages in Kenya may be due to the strong network of industry associations and institutions in the agricultural industry that support demonstration, collaboration, and competition spillovers. For example, the horticultural sector alone is supported by the Fresh Produce Exporters Association of Kenya (FPEAK), the Kenya Flower Council (KFC), the National

Task Force on Horticulture, and the Horticultural Crops Development Agency. Institutions that support both SME and large firm membership (such as FPEAK) are more likely to see spillovers in terms of demonstration, collaboration, and competition. For example, the KFC's membership is dominated by larger flower exporters and generally focuses on their specific concerns, but does not have many smaller members. The existence of a strong network of institutions and stakeholder forums was an important factor in the establishment of the KenyaGAP standard in 2007, as outlined in box 6.12.

Industry associations in the manufacturing industry also support demonstration, collaboration, and competition spillovers between foreign and domestic firms, albeit as part of wider efforts to support the manufacturing industry. In Kenya, organizations such as the Kenya Association of Manufacturers and the Kenya Private Sector Alliance facilitate linkages among firms.

The *level of agricultural research capacity* within public institutions also contributes to collaboration and demonstration spillovers in the agricultural sector, particularly where support institutions play a role in the dissemination of information and knowledge, such as in the coffee industry in Vietnam. In particular, the size, competence, and industry linkages of a country's public research institutions will influence the extent of spillovers from agricultural R&D.

As table 6.9 shows, Kenya and Ghana have more and better-resourced agricultural research institutions for R&D than Mozambique, where the main agricultural research agency, the Agricultural Research Institute of Mozambique, was only established in 2005. In Kenya, KARI accounts for 49 percent of total research spending. In Ghana, the main institutions are the nine agriculture-related research organizations within the Council for Scientific and Industrial Research (CSIR).

Kenya and Ghana's agricultural R&D institutions are funded primarily by national governments, although donor funding still plays a role in both countries. In Kenya, two nonprofit research foundations—the Tea Research Foundation (TRF) and Coffee Research Foundation—also exist. These are funded from commodity levies from industry stakeholders and the sales of goods and services, and promote industry-driven research. However, the

---

Box 6.12 KenyaGAP

KenyaGAP is a benchmark equivalent of the GlobalGAP standard demanded by U.K. and EU supermarkets. It was viewed as a way for Kenya to sustain its reputation as a producer of high-quality fresh produce in international markets. KenyaGAP customizes the GlobalGAP standards for Kenyan conditions by incorporating the needs of smallholders and outgrowers.

The government of Kenya mobilized stakeholders to participate in the development of the standard through the National Taskforce on Horticulture and the KenyaGAP Technical Committee. The Taskforce is a public-private sector initiative, including public bodies, exporters, growers associations, and international donors.

Table 6.9 Agricultural Research Capacity in Selected Countries

	Number of agricultural research institutions				Total agricultural research spending in 2008 (US$, millions)	Agricultural R&D spending as a percentage of agricultural GDP	Research staff (FTE)
	Total	Government funded	Higher education	Other			
Kenya	31	6	23	2	134.1	1.30	1,012
Ghana	27	12	15	0	77.6	0.90	537
Mozambique	5	2	3	0	16.1	0.38	263

Source: ASTI 2012.
Note: AgGDP = agricultural gross domestic product; FTE = full-time equivalent; R&D = research and development.

limited size of these organizations and the decline in commodity revenues have limited the scope of the research the organizations can conduct. For example, budget constraints have prevented the TRF from constructing a tea processing unit for research (Byerlee 2011).

In terms of private sector linkages within the agricultural research system, there are few linkages in Mozambique and Ghana, although Ghana has made some efforts to increase demand-driven research. These efforts include proposals for CSIR agencies to generate 30 percent of their funding target from the private sector, and a commitment to commercialization of research outputs. Ghana has also encouraged partnerships between private firms and NGOs and there are some examples of private sector collaboration with higher education and CSIR units in the areas of poultry feed and breeding stocks (Essegbey 2009). However, the impact of these efforts has been limited and private-public collaboration remains rare.

Conclusions

Of the three transmission channels—chain linkages; labor markets; and competition, demonstration, and collaboration effects—the supply chain appears to offer the greatest potential for spillovers. This is especially the case in the agricultural production segment of the value chain (somewhat less so for nonagricultural inputs), as domestic farmers are fundamental to foreign firms' operations in the sector. Following is an overview of the main conclusions drawn from this chapter.

Supply Chain Effects
Key Findings
1. FDI in Sub-Saharan Africa procures less locally, on average, than FDI in Vietnam. Levels of procurement vary across nature of investment, commodity, and country context.
2. For farmers, closer supply relationships lead to technical and management upgrading of suppliers through support from foreign investors.
3. More widely, foreign investment benefits farmers through access to new markets and reduced market risk.

4. Companies along the supply/distribution chain, other than farmers, also benefit through relationships with investors.

Factors Affecting Spillovers through Supply Chains
- Resource and market-seeking investors purchase more locally; market-seeking investors offer the best overall potential for spillovers.
- Value chain and commodity factors that contribute to greater local sourcing include: low capital and expertise requirements; need for a wide variety of smaller-scale inputs to complement a main product; need to reduce risks (for example, seasonality); low risks of side-selling; fluctuating demand.
- Tight regulatory environments (restricting access of FDI in certain sectors) and difficult access to land can lead to greater local sourcing, but may not necessarily develop competitive local producers.
- Greater formality and enforceable contracts facilitate local sourcing.
- Greater value-added processing in the country contributes substantially to greater local sourcing.
- Sophistication of domestic firms is among the most important factors for establishing local linkages, particularly in nonagricultural production.
- Market demands—especially certification and traceability standards—facilitate support for upgrading local suppliers; certification is also seen as a key tool for improving productivity.
- Collectives/cooperatives can be a facilitator or a barrier to spillovers, depending on their capacity and governance.
- Due to fragmentation, demonstration pilots (for technology, knowledge, and training) are critical.
- Complementing technical support with financial support (for example, prepayment) is important for sustainability and contributes to greater absorption; also important is combining technical training with business management training.
- Support through multistakeholder partnerships can be effective.
- The most effective public sector support comes from: (a) having high-quality research and extension services and institutions; (b) providing support and a regulatory push for standards and certification; and (c) facilitating access to finance.

Labor Market Effects
Key Findings
1. Most foreign investors aim to increase levels of local employment, and support skills development and knowledge of local staff, thereby enhancing the potential for spillovers.
2. A significant proportion of staff is reemployed at domestic firms, and a small number of staff starts businesses—resulting in some spillovers.
3. More widely, some foreign investors in Sub-Saharan Africa support skills development capacity through collaboration with education and training institutions.

Factors Affecting Spillovers through Labor Markets
- Local skills availability is by far the most important factor determining the extent of local labor linkages.
- Larger multinational firms and those with more decentralized operations tend to offer greater opportunities for local managerial and technical staff.
- Mobility between foreign and domestic firms may be limited by greater attractiveness of foreign firms.
- But the stronger and more sophisticated the base of domestic firms in a country, the greater the likelihood of labor circulation.
- Limited entrepreneurship, low access to finance, and difficult business environments constrain spillover through new ventures.
- Strong private sector organizations can facilitate industry collaboration with universities to address specific skills/experience gaps.

Competition, Demonstration, and Collaboration Effects
Key Finding:
1. Although collaboration in the industry remains at a relatively limited stage of development, it does present opportunities for upgrading both local firms and key sectoral institutions.

Factors Affecting Spillovers through Competition, Demonstration, and Collaboration
- Spillovers are more likely where domestic firms are larger and more technologically sophisticated.
- Representative institutions that include large and smaller firms as members increase the likelihood of spillovers by providing a forum for building social capital.
- High industry concentration limits collaboration potential.
- Market-seeking investors are most likely to invest in collaborative local R&D.
- Countries with strong public sector agricultural research capacity will help to enable collaboration with foreign investors.

Notes
1. Rupert Barnard, Michelle de Bruyn, Nick Kempson, and Philippa McLaren.
2. https://www.utzcertified.org.
3. http://www.rainforest-alliance.org/certification-verification.
4. http://cocobod.gh/index.php.
5. A failure to supply supermarkets with the required volumes stipulated in the contract can mean that the supermarkets can charge "loss of profit," which means the supplier is then liable for the retail price of the unfulfilled segment of the order.
6. http://www.globalhort.org/media/uploads/File/Videopercent20Conferences/VC3percent20KENYApercent20Positionpercent20Paper.pdf.

References

ASTI. 2012. *Global Assessment of Agricultural R&D Spending: Developing Countries Accelerate Investment*. Washington, DC: International Food Policy Research Institute.

Byerlee, D. 2011. "Producer Funding of R&D in Africa: An Underutilised Opportunity to Boost Commercial Agriculture." ASTI/IFPRI-FARA Conference Working Paper 4, International Food Policy Research Institute (IFPRI), and Forum for Agricultural Research in Africa (FARA). http://www.ifpri.org/publication/producer-funding-rd-africa-underutilized-opportunity-boost-commercial-agriculture.

Chamuene, A., N. Mahalambe, and O. Catine. 2010. *Summary of Production Trends, Cotton Research Status and Cotton IPM Experience in Mozambique*. International Cotton Advisory Committee. https://www.icac.org/tis/regional_networks/documents/africa_10/business_meeting/5_mozambique.pdf.

DRUM Commodities. 2012. "An African Tobacco Production Perspective: Executive Summary." *DRUM Commodities*. http://www.drumcommodities.com/assets/33/Tobacco_Project_Executive_Summary_April_2012.pdf.

Essegbey, G. 2009. "Ghana: Cassava, Cocoa, and Poultry." In *Agribusiness and Innovation Systems in Africa*, edited by K. Larsen, R. Kim, and F. Thues, 63–87. Washington, DC: World Bank.

Export Processing Zones Authority. 2005. *Tea and Coffee Industry in Kenya, 2005*. Export Processing Zones Authority, Nairobi. http://www.epzakenya.com/UserFiles/files/Beverages%20sector%20profile.pdf.

FAOSTAT. 2010. Database. *The Statistics Division of the Food and Agriculture Organization (FAO) of the United Nations*. http://faostat.fao.org/.

Gereffi, G. 1994. "The Organization of Buyer Driven Global Commodity Chains: How U.S. Retailers Shape Overseas Production Networks." *Commodity Chains and Global Capitalism*, edited by G. Gereffi and M. Korzeniewicz, 95–122. Westport, CT: Praeger.

Gereffi, G., K. Fernandez-Stark, and P. Psilos. 2011. *Skills for Upgrading: Workforce Development and Global Value Chains in Developing Countries*. Duke University, Center on Globalization, Governance & Competitiveness. Durham, NC: Duke University.

Gereffi, G., and J. Lee. 2009. *A Global Value Chain Approach to Food Safety and Quality Standards*. Report prepared for the Global Health Diplomacy for Chronic Disease Protection Working Paper Series, February 4.

Gibbon, P., and S. Ponte. 2005. "Trading Down: Africa, Value Chains, and the Global Economy." *The Journal of Modern African Studies* 45 (1): 178–79.

Nyabiage, J. 2010. "Kenya: Cadbury Shuts its Chocolate Division." *All Africa*, November 15. http://allafrica.com/stories/201011160089.html.

Paglietti, L., and R. Sabrie. 2012. "Outgrower Schemes: Advantages of Different Business Models for Sustainable Crop Intensification: Ghana Case Studies." FAO Investment Centre—Learning from Investment Practices. http://www.fao.org/fileadmin/user_upload/tci/docs/LN1-Ghana_Outgrower%20schemes.pdf.

Reardon, T., S. Henson, and J. Berdegué. 2007. "'Proactive Fast-Tracking' Diffusion of Supermarkets in Developing Countries: Implications for Market Institutions and Trade." *Journal of Economic Geography* 7 (4): 399–431.

VECO. 2011. *Increased Incomes for Indonesian Cocoa Farmers in Sustainable Markets: NGO–Private Sector Cooperation on Sulawesi Island.* VECO, Leuven, Belgium. http://www.veco-ngo.org/sites/www.veco-ngo.org/files/blog/bijlage/indonesia-cocoa-case.pdf.

Vermeulen, S., and N. Goad. 2006. *Towards Better Practice in Smallholder Palm Oil Production.* Natural Resources Issues Series No. 5. London: International Institute for Environment and Development. Available from IIED on request.

Zen, Z., Z. Lubis, and S. Edyono. 2005. *Social Mapping and Identification of Strategic Issues in Communities Surrounding PT X at Kabupaten Y in Sumatra.* Medan: Universitas Muhammadiyah.

CHAPTER 7

Sector Case Study: Apparel

Cornelia Staritz and Stacey Frederick[1]

Abstract

At the beginning of the 2000s, the introduction of the U.S. African Growth and Opportunity Act, combined with Multi-Fibre Arrangement quotas, contributed to a boom in foreign direct investment (FDI) in the Sub-Saharan African (SSA) apparel sector, leading to major growth in production, exports, and jobs. The possibility of exploiting the spillover potential of this FDI raised significant hopes of developing a locally embedded, competitive SSA apparel export industry. Yet more than a decade later, there has been very little progress made in reaching this objective, outside of Mauritius (and South Africa where FDI did not play a prominent role). Despite significant investments to attract FDI through building export processing zones and offering fiscal incentives, virtually no locally owned apparel firms are exporting or subcontracting, local value added remains low, local participation in management is limited, and domestic suppliers are almost absent in core and even most noncore inputs. This case study explores the level and nature of spillovers in three of the leading SSA apparel-exporting countries—Kenya, Lesotho, and Swaziland—and endeavors to understand the factors constraining them.

We find across all countries FDI strategies that severely limit spillover potential from the start, including a concentration in low value–added activities, external control of sourcing, and reliance on expatriates in managerial and technical positions. This is aggravated by a weak domestic absorptive capacity (weak skills development, nonexisting or inadequate domestic training institutes), barriers in the domestic business climate, ineffective policies to support local small and medium enterprises, and a missing local entrepreneurial response.

The Apparel Global Value Chain[2]

Introduction to the Apparel Global Value Chain

Apparel is one of the largest export sectors in the world and has become one of the most globalized. In recent decades, developing countries have come to dominate world trade in the sector, expanding their share from around 25 percent in the 1960s to more than 80 percent today. Indeed, the apparel sector has long played a central role in the industrialization process of low-income countries (LICs), absorbing large numbers of unskilled, mostly female, workers and provided upgrading opportunities into higher value–added activities within and across sectors.

As in many other sectors, production and trade in the apparel sector are organized in global value chains (GVCs) where production of components and assembly into final products is carried out via inter-firm networks on a global scale. To simplify the many activities in the chain, activities can be separated into the apparel supply chain and the apparel value chain. The supply chain focuses on the physical transformation and transportation of raw materials to final products, and the value chain focuses on activities that add economic value to products at each stage, but are not necessarily manufacturing or logistics related. Together, these make up the GVC for apparel.

The apparel supply chain can be roughly divided into four stages that are intertwined with the textile sector (figure 7.1): (a) raw material supply, including natural (for example, cotton, wool, silk, and flax) and synthetic or man-made fibers (for example, polyester, nylon, and acrylic); (b) yarn and fabric production and finishing (for example, dying, printing, cutting); (c) production (for example, clothing, home furnishing, and industrial and technical products); and (d) distribution and sales at the wholesale and retail levels.

A large part of apparel production—including cutting, sewing, and finishing activities—remains labor intensive, has low start-up and fixed costs, and requires simple technology. These characteristics have encouraged the move to low-cost locations, mainly in developing countries. In contrast, textile (yarn and

Figure 7.1 Apparel Supply and Value Chain

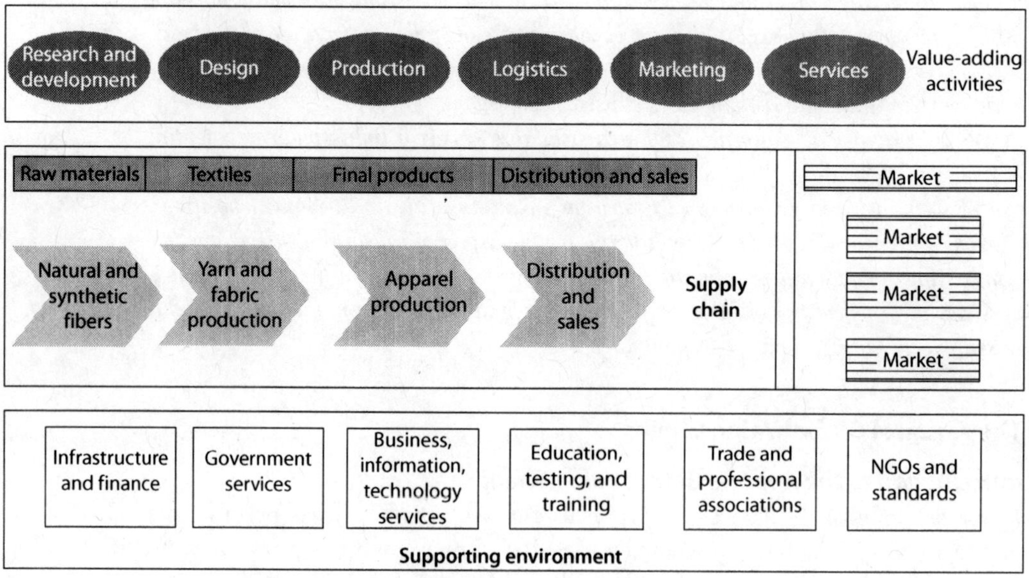

Source: Frederick 2010.
Note: NGO = nongovernmental organization.

fabric) production is more capital and scale intensive, demands higher skills, and has partly remained in developed countries or shifted toward middle-income countries. Within the apparel sector, there are two segments, woven and knitted apparel. The value chains of these two types of apparel are quite distinct—they use different types of yarn, fabric, machinery, and manufacturing technology, and they also differ in terms of labor use as they have different capital-intensity and skill requirements. They also differ with regard to vertical integration; knitted apparel producers are more often vertically integrated into fabric and even yarn production. In the woven segment, investments into fabric weaving are substantially larger and are generally set up in independent weaving mills.

In addition to the physical manufacturing-related steps in the textile-apparel supply chain, there is also a series of "intangible" activities that add economic value to apparel products. The apparel value chain consists of seven main value-adding activities including: product development, design, textile sourcing, apparel manufacturing, distribution, branding, and retail. These activities link producers to consumers and are controlled by a combination of lead firms, intermediaries, and apparel manufacturers.

Apparel represents a classic example of a buyer-driven value chain.[3] Buyer-driven value chains are common in labor-intensive, consumer goods industries such as apparel, footwear, toys, and consumer electronics. They are characterized by decentralized, globally dispersed production networks, coordinated by lead firms that control high value–added activities (for example, design and branding) and often outsource the (largely standardized) manufacturing process to a global network of suppliers (Gereffi 1994, 1999; Gereffi and Memedovic 2003).

Main Actors in the Apparel Value Chain
Lead Firms

The strategies of lead firms, in particular their global sourcing policies, importantly shape production and trade patterns in the apparel sector. There are three main types of lead firms in the apparel GVC: brand marketers, brand manufacturers, and retailers (mass market and specialty) (figure 7.2).

- *Brand manufacturers* (such as VF Corporation, Hanesbrands, Fruit of the Loom, and Levi's) own apparel manufacturing plants, own textile plants or coordinate textile sourcing, and control the marketing and branding activities in the chain. Their production networks are often set up in countries with reciprocal trade agreements where these firms have an important role as foreign investors. In the 1970s and 80s, the brand manufacturer category was much more significant, but has been on the decline over the last two decades as manufacturers have started outsourcing production-related activities to focus on higher-value segments of the chain such as retail and marketing.

- *Brand marketers* (such as Nike, Reebok, Liz Claiborne, and Ralph Lauren) control the branding and marketing functions, but do not own manufacturing

Figure 7.2 Types of Lead Firms in Apparel GVCs

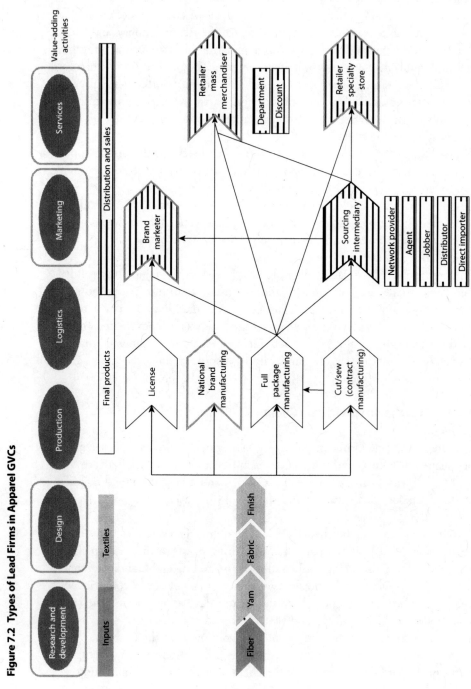

Source: Frederick 2010.
Note: GVC = global value chain.

facilities. Brand marketers are either former brand manufacturers that moved out of production or firms that emerged as apparel designers.

- *Retailers* are involved with the branding and marketing of product lines developed for and sold only via their retail locations. Mass market and department store retailers (such as Wal-Mart, Target, JC Penney, Marks & Spencer, and Tesco) are separated from specialty stores (such as Gap, Limited, H&M, Mango, and New Look) because the latter only sell apparel-related merchandise under their brand names. Retailers do not own manufacturing facilities, but work directly with an apparel manufacturer or with a sourcing agent, who coordinates the supply chain.

Sourcing decisions of lead firms and related production relocations traditionally have been motivated primarily by labor cost differentials as well as market access. More recently, however, sourcing policies have become more complex. There is greater demand for lead times and flexibility (linked to trends toward lean retailing and "fast fashion"), compliance with labor, social, and environmental standards, and emphasis on nonmanufacturing capabilities of firms (for example, inventory management, financing, logistics, and so forth).

Intermediaries and Triangular Manufacturing Networks

Intermediaries such as importers, exporters, agents, and trading houses play a central role in apparel GVCs and have provided key links between buyers and suppliers. The role of intermediaries has evolved: buyers have tried to bypass them to reduce costs and increase control over the supply chain, while suppliers have built up capabilities that allowed them to deal directly with buyers. However, smaller buyers still rely on traditional intermediaries such as buying agents (see, for example, Palpacuer, Gibbon, and Thomen 2005 for Europe). In the 1990s, a new type of intermediary evolved that has played an important and increasing role in apparel GVCs. Large full-package manufacturers, in particular in East Asia, have developed from producers to intermediaries and organize far-flung multinational production and sourcing networks (Appelbaum 2008). More recently, large manufacturers in other Asian countries such as China, India, Malaysia, Singapore, and Sri Lanka as well as the Middle East have also developed multinational manufacturing networks. These multinational producers have become an important source of foreign direct investment (FDI) in apparel GVCs.

Suppliers

Suppliers and subcontractors in the apparel sector can be classified along a spectrum based on their functional capabilities—that is the range of activities they carry out and the relative value they add. These can be summarized as follows:

- *Assembly and CMT (cut-make-trim):* The apparel manufacturer is responsible for sewing the apparel and may be responsible for cutting the fabric and providing simple trim (buttons, zippers, and so forth). The buyer provides product

specifications and the fabric. The apparel factory is paid a processing fee rather than a price for the product.
- *Full Package Provider and Free on Board:* The apparel manufacturer purchases (or produces) textile inputs and provides all production services, finishing, and packaging for delivery to the retail outlet. The customer provides the design and often specifies textile suppliers.
- *Full Package with Design and Original Design Manufacturing:* The apparel supplier is involved in the design and product development process, including the approval of samples and the selection, purchase, and production of required materials. The apparel supplier is also responsible for coordinating sourcing activities.
- *Original Brand Manufacturing (OBM):* The apparel supplier is responsible for branding and marketing of the final products. The apparel firm may do these activities on a contract basis on behalf of another lead firm, or it can mark the transition from apparel supplier to a lead firm, typically in domestic or regional markets.

Apparel-producing countries are often categorized by the functional capabilities of the majority of apparel manufacturing firms within the country (despite important variations within countries) (table 7.1). Caribbean and Sub-Saharan

Table 7.1 Firm Characteristics and Needs by Functional Category

Category	Characteristics and needs	
CMT	*Basic products:*	*Complex products:*
	Low minimum wages	Access to employees trained in quality control
	Low operating costs	Employees with complex product knowledge
	Trade preferences	Available capital to purchase finishing equipment
		Access to Internet to communicate with buyers
FOB: sourcing	Available capital to purchase textiles and trim	
	Knowledge of and access to potential suppliers	
	Employees with knowledge of textile products and trim, and logistics and inventory management	
	Good transportation infrastructure (roads and ports) and presence of logistics and transportation providers	
	Reliable Internet infrastructure to communicate with suppliers	
FOB: textile production	Available capital to invest in textile machinery	
	Access to reliable and inexpensive energy	
	Employees with production-related skills related to textile products and machinery operation	
	Critical mass of potential buyers in the vicinity and/or easy access to ports	
ODM	Employees with product development skills and knowledge of size standards	
	Available capital to purchase computer-aided design (CAD) equipment	
	Presence of local designers and access to workers trained in CAD	
OBM & retail	Access to retailers (OBM); access to local and regional ODM and OBM apparel suppliers (retail)	
	Available capital to invest in market research and capital to invest in advertising	
	Access to workers with research, marketing, and management skills	

Note: CMT = cut-make-trim; FOB = free on board; ODM = original design manufacturing; OBM = original brand manufacturing.

African (SSA) countries (with the exception of South Africa and Mauritius) as well as Cambodia are typically limited to CMT capabilities, usually focusing on low-cost volume products. An increasing share of apparel manufacturers in Bangladesh, Indonesia, Mexico, and Vietnam fall into the full package category. China, India, and Turkey have apparel exporters with full package and design capabilities, but also OBM capabilities in their home markets.

Table 7.1 describes apparel producers by their characteristics and needs by their functional category in the value chain.

FDI in the Apparel Global Value Chain

Global Trade and Investment Context

Besides the crucial importance of organizational dynamics, institutional and regulatory factors decisively influence global production and trade patterns and upgrading prospects in apparel GVCs. The apparel sector has been one of the most trade-regulated manufacturing activities in the global economy. Until 2005, textile and apparel trade had been governed by a system of quantitative restrictions for more than 40 years under the Multi-Fibre Arrangement (MFA). The objective of the MFA was to protect the major import markets (Europe, the United States, and Canada) by imposing quotas on the volume of imports and to allow those countries to restructure their sectors before opening up to competition. However, the result was to promote the spread of apparel production in developing countries. When manufacturers (mostly from Japan; the Republic of Korea; Hong Kong SAR, China; Taiwan, China; and later China) reached quota limits in their home economies, they searched for producer countries with underutilized quotas or for countries with no quota to set up apparel plants there or source from existing apparel firms.

The Agreement on Textiles and Clothing signed in 1994, committed to phase out the MFA by the end of 2004 by eliminating all quotas on textile and apparel trade between World Trade Organization members (although in practice some temporary safeguards remained in place through 2008). This had profound implications on the structure of the global apparel sector. China's export share rose dramatically from 28 percent in 2004 to almost 43 percent by 2010 (table 7.2). China and the rest of the so-called "Asian 12"[4] now account for 63 percent of global apparel trade. By contrast, regional developing country suppliers to the main import markets of the United States and European Union (EU)—including those in Eastern Europe, North Africa, and Latin America and the Caribbean—have generally lost share over this period, although they still remain relevant in the context of increasing demand for flexible production and logistics. Previously dominant Asian export locations such as Hong Kong SAR, China; Republic of Korea; the Philippines; and Taiwan, China, have also steadily lost share over the past decade. Finally, SSA countries still hardly register in terms of global trade—their aggregate share of the global market is below 1 percent.

While quotas are no longer a feature of apparel trade and investment, tariffs still play a central role in all important global markets. Average most

Table 7.2 Top 15 Apparel Exporting Countries, 2010

Country and region	Value (US$ million)			World share (percent)		
	2000	2005	2010	2000	2005	2010
World	193,669	268,431	326,254			
China	48,019	89,890	139,900	24.8	33.5	42.9
EU-15	33,980	47,598	51,898	17.5	17.7	15.9
Bangladesh	4,862	8,038	16,620	2.5	3.0	5.1
Turkey	6,711	12,942	14,759	3.5	4.8	4.5
India	5,131	9,476	12,877	2.6	3.5	3.9
Vietnam	n.a.	4,739	10,953	n.a.	1.8	3.4
Indonesia	4,675	5,679	7,894	2.4	2.1	2.4
Mexico	8,924	6,683	4,199	4.6	2.5	1.3
Cambodia	n.a.	n.a.	4,184	n.a.	n.a.	1.3
Morocco	n.a.	3,331	3,765	n.a.	1.2	1.2
Tunisia	2,645	3,478	3,730	1.4	1.3	1.1
Sri Lanka	n.a.	3,083	3,729	n.a.	1.1	1.1
Thailand	3,672	3,862	3,725	1.9	1.4	1.1
Pakistan	n.a.	n.a.	3,679	n.a.	n.a.	1.1
Romania	2,737	5,177	3,327	1.4	1.9	1.0
Top 15	147,007	216,185	285,238	75.9	80.5	87.4

Source: United Nations Commodity Trade Statistics Database (UN Comtrade), retrieved 6/2/2012.
Note: Apparel represented by HS codes 61 and HS 62. Top 15 values are by year. n.a. = not applicable (country not in top 15 in given year).

favored nation tariffs on imports of textiles are 6.7 percent for the EU and 7.5 percent for the United States; for apparel they are 11.5 percent and 11.4 percent respectively. Tariffs vary considerably for different product categories, with peaks of 12 percent in the EU and 32 percent in the United States. Given the significance of tariffs in a competitive and relatively low-margin industry, preferential market access has a substantial impact on global production and trade patterns, driving investment location decisions in the sector. The early 2000s saw a proliferation of such agreements, mainly designed to give preferences to LICs. In the context of SSA, these include, most notably, the introduction of the African Growth and Opportunity Act (AGOA) by the United States in May 2000 and Everything But Arms in the EU in March 2001.

Preferential market access in these agreements is governed by rules of origin (ROO). For example, ROO may require a certain level of local value added, or specify the country or region of origin of key inputs like fabric. These ROO have a crucial impact on outcomes, both in terms of FDI and the scope for local sourcing and spillovers.

FDI Development and Recent Trends

Starting in the 1970s, large domestic firms in the United States, Europe, and Japan began to shift labor-intensive activities to lower-cost countries as part of a global restructuring of their operations (McCormick and Schmitz 2001). Two modes of apparel distribution networks emerged from this:

- *Production models*: The buyer (typically a brand manufacturer) is involved in the production process, either through ownership (domestic or foreign) or by supplying inputs to production. These networks were typically regional (Mexico and the Caribbean Basin for the United States; Central and Eastern Europe and North Africa for Europe).

- *Sourcing models*: The buyer (typically a brand market or retailer) does not own or maintain involvement in production, but merely sources or purchases a product from the final manufacturer or an agent (typically used by brand marketers and retailers) (Frederick 2010); these networks were typically global in scale. In this model, buyers work with networks of contract manufacturers or full package suppliers. These "manufacturers without factories" initially sourced apparel primarily from East Asia, mostly from Hong Kong SAR, China; Korea; and Taiwan, China. The reason was that these manufacturers could produce orders for finished apparel to the buyers' specifications (Bair 2006).

Three major trends have shaped FDI patterns in apparel GVCs over the last two decades:

- *Shift in lead firms from manufacturers to marketers and retailers*: Over the past 20 years there has been a significant power shift in the retail and marketing portion of the apparel value chain from producers to buyers (especially retailers). Retailers now dominate imports and sourcing decisions. The implication is that the sourcing model has become much more important compared with the FDI-based production model, contributing to declining direct FDI from the United States and EU brand manufacturers and increasing the importance of global contract manufacturers.

- *Rise of FDI from Asian-based multinationals*: Apparel firms, in particular in the "Big Three" (Hong Kong SAR, China; Korea; and Taiwan, China), developed from producers to intermediaries by organizing far-flung multinational production and sourcing networks. This shift resulted from a combination of rising cost pressures in their home economies and the need to get around MFA quota limitations. These firms established production networks in the Asian region in the 1970s and 1980s, and extended them to Latin America and SSA in the 1990s. In doing so, they changed dramatically the geographical sourcing strategies of global buyers.

- *Increasing role of regional investments*: Regional manufacturing networks have emerged in SSA, South Asia, and Southeast Asia. In SSA, particularly manufacturers from Mauritius have invested in Madagascar and to a lesser extent in Ghana and Senegal. South African firms have invested in Lesotho.

In South Asia, large manufacturers in India and Sri Lanka have built regional production networks by investing in lower value–added production activities in Bangladesh and Pakistan. Sri Lankan manufacturers have also invested in India, particularly in textile mills. In Southeast Asia, particularly Malaysian and Thai manufacturers have invested in apparel plants in Cambodia and Lao People's Democratic Republic to take advantage of lower labor costs and preferential market access to the EU. These investments have increased in the context of regional integration efforts.

FDI Spillover Channels and Challenges in Apparel GVCs

This section identifies the main FDI spillover channels in the apparel GVC, and gives an overview of the nature of FDI spillovers and challenges in achieving spillovers in apparel GVCs globally. There are several general characteristics of apparel sector FDI that tend to shape spillover potential:

- Apparel FDI is largely efficiency seeking and export oriented, particularly in LICs. Given the labor-intensive production process, labor cost is a crucial factor in FDI decisions. Trade policy factors are also important, in particular preferential market access. More recently, sourcing and FDI decisions have become more complex, taking into account lead time and flexibility, compliance, and broader nonmanufacturing capabilities.
- Given the labor-intensive and relatively low-tech nature of the apparel sector, FDI tends to create a large number of jobs, primarily for unskilled or semiskilled positions, which are often filled by female workers. Apparel FDI also provides relatively few high-skilled jobs and limited high-tech or knowledge-intensive processes or equipment.
- Investment costs are relatively low in the apparel sector given its labor intensity, low start-up and fixed costs, and simple technology. On the contrary, investments in backward linkages to the textile sector require substantially higher initial capital costs.
- Apparel FDI tends to be greenfield in nature and on a larger scale than local investment. But compared to other sectors, for example, mining, its scale tends to be substantially smaller.

Supply Chain Linkages

There are three main types of supply chain linkages with foreign-owned apparel firms: inputs and services for apparel production (backward linkages), subcontracting of apparel production, and local firms using the output of foreign-owned apparel firms (forward linkages). In the apparel sector, backward linkages and subcontracting are generally more important than forward linkages as apparel is a final product, and FDI, particularly in LICs, is generally export oriented.

There are several categories of industry-specific backward supply chain linkages and subcontracting opportunities in the apparel industry. These include: (a) direct raw material inputs (for example, fabric and yarn); (b) apparel trim and accessories (for example, buttons, zippers, thread, elastic, labels, and hangers);

(c) nonessential inputs such as packaging (for example, cartons and poly bags); (d) capital equipment and machinery parts manufacturers or suppliers; and (e) subcontractors that perform a portion of assembly or finishing activities on behalf of another firm (for example, sewing, embroidering, or screen printing). A sixth, general category also includes broad services applicable to a range of industries such as transportation, logistics, catering, information technology (IT), construction, cleaning, security, human resource, and training.

The most important backward linkage is to the textile sector, as fabrics are the most expensive input into apparel production and the quality of the textiles is directly related to the final product's quality. However, in contrast to apparel production, *textile production* is more capital-, skill-, and scale-intensive, which is a challenge for the establishment of backward linkages.[5] A certain minimum size of the apparel industry, locally or regionally, is a requirement for local or foreign investment into backward linkages, particularly in the woven segment. Hence, backward supply chain linkages into textiles are more likely to occur if major foreign firms invest in the country or region or if major buyers purchase a significant share of apparel from the country.

Besides these issues related to the local business environment and capabilities, sourcing policies of buyers and foreign investors are crucial for determining the extent of backward linkages into textile production. Many buyers nominate textile mills that have to be used to source textile inputs for their orders. The motivation behind this is threefold: First, many buyers have long-established relationships with textile producers that fulfill their specifications with regard to quality, reliability, and costs. Second, nominating textile mills helps with quality control, particularly with dyed textiles, and helps to assure that inputs used by different apparel manufacturing facilities have the same color. Third, buyers often have more purchasing power than the apparel manufacturers and can negotiate better rates.

In addition, the head offices of foreign investors are generally in charge of input sourcing and they may have their own textile mills from which they source inputs for all their apparel plants on a global scale. Brand manufacturers often own textile plants in the United States and European countries, which they use as input suppliers for their offshore apparel plants. However, more recently, several brand manufactures have also relocated their textile production to nearby regional supplier countries. Multinational producers often also own their own textile plants in their home country or in core economies in East Asia and China. To a certain extent, multinational producers have also relocated textile mills closer to their main sourcing countries. For instance, Nien Hsing established a denim fabric mill in Lesotho to supply its apparel plants in Lesotho and other SSA countries. Such co-location strategies of global suppliers increase FDI and, due to the capital-intensity of textiles, make it more embedded; but these strategies also make it more difficult for local firms to enter sourcing networks and build supply linkages.

Given the prevalent global sourcing strategies with regard to textiles, foreign investors also tend to source *other inputs* such as accessories close to the textile mills and send all inputs together in a box to their global apparel plants. However, multinational trims suppliers such as YKK also have plants on a regional basis. For instance, YKK owns a plant in Swaziland to supply the SSA market with zippers. In another case, Coats Threads invested in a polyester thread dyeing facility in Madagascar (De Coster 2002). Again, such co-location strategies further FDI but also make it difficult for local firms to establish linkages. Low-value and often nonapparel-related inputs such as labels and packaging material are more broadly sourced locally.

Specialized core *services*, including logistics, shipping, and IT, tend to be supplied by providers with a global or regional footprint. For instance, large globally operating service providers tend to have an office in South Africa from which they provide services to the SSA market. Nonspecialized services, including catering, local transportation, construction, cleaning, and certain human resource and training services, are more likely to be sourced from local firms. Services such as embroidery, laundry, and printing can be either fulfilled in-house or outsourced to independent firms. In countries such as Bangladesh, Mauritius, and Sri Lanka, these service activities were an entry for local firms.

Subcontracting of CMT activities is another important linkage channel and entry point to exporting for locally owned firms. Given the difficulties in establishing direct relationships with buyers and sourcing networks, fulfilling subcontracting work for foreign-owned firms offered entry and experiences in export-oriented apparel production. Further, foreign-owned firms may support process upgrading or even product upgrading by giving assistance on production setup productivity improvements, and quality standards.

Labor Markets and Human Capital

Apparel FDI creates employment on a large scale, primarily for low- and semi-skilled, often female, workers. Due to the low-skill nature of work, spillover potential is limited for production workers. In higher-skilled technical and management positions, the use of expatriate workers is widespread among foreign-owned companies, although there are important differences among companies.

The lack of established local manufacturing firms in many LICs, particularly in SSA, and/or the limited interaction between foreign and local firms may constrain labor mobility and spillovers between foreign and local firms. The limited interaction of local and foreign firms may be related to very different production models and market segments of these firms. Foreign firms tend to focus on large-run export production while local firms, if they exist, produce largely for local or regional markets, often on a smaller scale or even informally, as is the case in many LICs.

Competition and Demonstration Effects

The extent and potential for technology and knowledge spillovers through demonstration depends first on the extent of knowledge transfer from head offices to foreign affiliates, and second on the degree of interaction between foreign and local firms. If foreign investors only relocate specific, low-value, production-related functions to their foreign plants, as is the case with many apparel investments, the spillover potential is limited from the onset.

Spillovers are further dependent on the existence of local firms that can absorb knowledge and technology from foreign affiliates and by interaction between foreign and local firms. As discussed, this can be limited in the context of LICs, particularly in SSA. Further, demonstration spillovers through imitation or reverse engineering may be limited by the existence of segregated production locations of local and foreign (often export-oriented) firms, which reduces interactions. In particular, the importance of export processing zones (EPZs) or similar zones in the apparel sector may be a barrier to spillovers to the extent that local firms, related to their smaller scale, have a limited presence in EPZs.

The Context for Spillovers in Sub-Saharan Africa

Apparel Sector Development in Sub-Saharan Africa

Over the past decade, several Sub-Saharan Africa (SSA) countries have developed export-oriented apparel sectors, in particular Kenya, Lesotho, Madagascar, and Swaziland, as well as Mauritius where the process started in the 1970s. This took place, first, within a policy framework of "export-led growth" as governments hoped that the sector would play a central role in (starting) the industrialization process as it did in other countries. Second, apparel exports grew in light of MFA quota restrictions on large Asian producing countries and based on agreements securing preferential market access for developed countries, in particular AGOA. SSA apparel exports increased from US$1.3 billion in 1997 to US$3.2 billion in 2004 (table 7.3) and dramatically changed composition. Exports to the EU stagnated while those to the United States more than doubled, peaking at US$1.8 billion for all SSA AGOA beneficiaries in 2004. The share of SSA apparel exports in global apparel exports increased to 1.3 percent in 2004; in the United States, the region's share reached 2.6 percent in 2004.

However, since around 2004 the industry has declined quite drastically in terms of production, exports, employment, and number of firms in all of the main SSA apparel-exporter countries. The total value of SSA apparel exports decreased by 20 percent from 2004 to 2008 (table 7.3); exports to the United States declined by one-third over this period (table 7.4). Since 2008–09 there have been further declines in exports. They are associated, first, with the abolition

Table 7.3 Sub-Saharan African Apparel Exports to the World

	1995	1997	2000	2002	2004	2005	2006	2007	2008	2009	2010
Total value of exports (US$, millions)	1,137	1,293	2,089	2,288	3,235	2,796	2,765	2,995	2,826	2,271	2,040
Growth rate (%)	13.2	8.0	46.1	1.6	12.4	−13.6	−1.1	8.3	−5.7	−19.6	−10.2
Share of world (%)	0.7	0.8	1.1	1.1	1.3	1.0	1.0	0.9	0.8	0.8	0.6

Source: United Nations Commodity Trade Statistics Database (UN Comtrade), retrieved 5/2/2012.
Note: Apparel represents HS(92) codes 61+62; exports represent world imports.

Table 7.4 Top 10 Sub-Saharan African Apparel Exporters to the United States by Year

	Value ($US, millions)						Share of SSA total (percent)					
Exporter	2000	2002	2004	2007	2010	2011	2000	2002	2004	2007	2010	2011
SSA total	748	1,098	1,757	1,293	790	904						
Lesotho	140	321	456	384	281	315	18.7	29.3	25.9	29.7	35.5	34.9
Kenya	44	125	277	248	202	261	5.9	11.4	15.8	19.2	25.5	28.8
Mauritius	245	254	226	115	120	157	32.7	23.2	12.9	8.9	15.1	17.3
Swaziland	32	89	179	135	93	77	4.3	8.1	10.2	10.5	11.8	8.5
Madagascar	110	89	323	290	55	40	14.7	8.1	18.4	22.4	6.9	4.4
Botswana	8	6	20	31	12	15	1.1	0.6	1.2	2.4	1.5	1.7
Malawi	7	11	27	20	10	13	1.0	1.0	1.5	1.5	1.3	1.5
Ethiopia	n.a.	n.a.	n.a.	n.a.	7	10	n.a.	n.a.	n.a.	n.a.	0.8	1.1
South Africa	142	181	141	24	6	7	18.9	16.4	8.0	1.8	0.8	0.7
Tanzania	n.a.	n.a.	n.a.	n.a.	2	5	n.a.	n.a.	n.a.	n.a.	0.2	0.6
Top 10	746	1,092	1,735	1,283	788	900	99.8	99.5	98.8	99.2	99.6	99.5

Source: United States International Trade Commission (USITC), General Customs Value.
Note: Apparel represents HS codes 61+62; exports represented by U.S. imports. n.a. = not applicable; SSA = Sub-Saharan Africa.

of safeguard quotas on U.S. and EU imports of apparel from China, and, more importantly, with reductions in U.S. and EU apparel imports as a result of the global economic crisis. The total value of SSA apparel exports declined by 5.7 percent in 2008, 19.6 percent in 2009, and 10.2 percent in 2010 (table 7.3). The net result is that the region's global market share fell to 0.6 percent, below even its 1995 level. Despite this, apparel remains the most important manufactured export from SSA (Kaplinsky and Morris 2008).

FDI in Kenya, Lesotho, and Swaziland

In the three SSA countries analyzed, the majority of export-oriented apparel firms are foreign owned. In Lesotho, there are currently 31 apparel firms (37 plants) and one textile mill operating in the formal apparel and textile manufacturing industry; 16 firms (21 plants) are Asian-owned (12 Taiwanese and 4 [mainland] Chinese), 14 South African–owned (15 plants), and 1 firm is Mauritian. There are no locally owned firms.[6] In Swaziland, there are currently 13 apparel firms (16 plants) and 3 textile mills (owned by two foreign-owned firms; one Taiwanese and one South African) operating in the formal apparel and textile manufacturing industry. Of the apparel firms, nine firms are Taiwanese-owned, three South African–owned, and one is locally owned. In Kenya, there are 18 apparel firms operating in EPZs—6 firms are Indian-owned with 2 having head offices in Dubai; 4 are from Taiwan, China; 4 are from China; 1 is from Hong Kong SAR, China; 2 have joint ownership structures; and 1 is locally owned. Interviews and surveys were conducted with several of these foreign-owned firms in each country; 15 in Kenya, 13 in Lesotho, and 11 in Swaziland. This sample represents over 80 percent of the foreign-owned, export-oriented EPZ firms in Kenya, all but one foreign-owned firm in Swaziland, and 42 percent of foreign-owned firms in Lesotho. Figure 7.3a provides an overview of the

Figure 7.3 Nationality and Export Destination of Foreign-Owned Apparel Firms

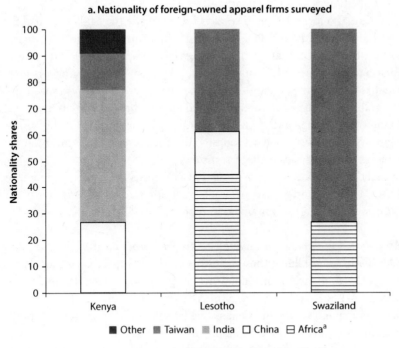

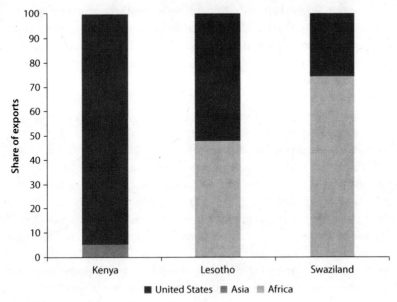

Source: Surveys and interviews of 39 foreign-owned firms: Kenya = 15, Lesotho = 13, Swaziland = 11.
a. Africa includes South Africa in Lesotho and Swaziland and one Mauritian investor in Lesotho.

ownership structures of the foreign firms interviewed in the three countries; Figure 7.3b shows export destinations.

There are also foreign-owned apparel firms outside the EPZs in Kenya that are largely Indian-owned (and locally owned). In contrast to the EPZ firms that export to the United States, these foreign-owned firms focus on apparel and often also textile production for the local and regional market. Surveys were also conducted with some of these firms; however there are very limited linkages between these two types of foreign-owned apparel firms, to be discussed further. In Lesotho and Swaziland, there are no other formal apparel firms besides the foreign-owned export-oriented ones; only very small-scale, informal, locally owned, and local market–focused firms exist.

Broadly, three types of export-oriented FDI firms exist in the apparel sectors in Kenya, Lesotho, and Swaziland, each with different spillover potential. These are discussed here and summarized in table 7.5.

- *Asian-based multinational producers*: Largely from Taiwan, China; Hong Kong SAR, China; and China; these investors exist in all three countries. They have head offices in their home countries and organize triangular manufacturing networks. Local activities are limited to manufacturing. Head offices hold decision-making power and manage higher-value functions such as input sourcing and financing, product development and design, sales and marketing, logistics, and buyer relationships. The strategy of these firms is global: exporting long-run, basic products almost exclusively to the U.S. market with manufacturing plants in different regions. Expatriates generally have an important role in management. General managers of these firms are generally employees and not owners of the firms. Multinational producers often own textile mills in other countries that are used for their apparel manufacturing plants. They also tend to source inputs on a global scale for their globally dispersed manufacturing plants to get better prices and secure conformity. The FDI spillover potential of these firms is limited given their specific global production and sourcing model.

- *More locally embedded Asian investors*: These are typically single operations that have their head offices and sourcing and sales competencies in Africa. In Kenya, most Indian investors can be labeled as more embedded foreign investors; in Lesotho and Swaziland, this group includes some more-embedded Taiwanese, Chinese, and one Mauritian investor. In these firms the general manager is generally also the owner (or at least part owner) of the firm and not an employee. Most own or work with sourcing offices in India; Taiwan, China; and/or Hong Kong SAR, China; or also the United States to get orders and have contacts with buyers as well as to source inputs. These firms overwhelmingly export to the U.S. market, but in Lesotho and Swaziland they have also tried to export to the South African market. The FDI spillover potential of these firms is generally higher as they are not part of tightly organized global production networks. They have a more fluid division of labor and local sourcing, as well as merchandising and sales competencies. However, these firms generally have no close

Table 7.5 Types of FDI and Spillover Potential in Kenya, Lesotho, and Swaziland

Characteristics and spillover potential	Type of FDI			
	Multinational producers	More embedded multinational investors		Regional investors
Nationality	China; Hong Kong SAR, China; Taiwan, China	China; India; Mauritius; Taiwan, China		South Africa
Importance	Kenya: 8; Lesotho: 10 and 15; Swaziland: 2	Kenya: 8; Lesotho: 6; Swaziland: 6		Kenya: 0; Lesotho: 16; Swaziland: 3
Investor motivation	MFA quota, AGOA, investment incentives	MFA quota, AGOA, investment incentives, industry contacts		Labor costs, investment incentives, Southern African Customs Union (SACU)
Functional capabilities	CMT, higher-value functions, and decision making abroad	CMT with some FOB, local decision making		Largely CMT, higher-value functions, and decision making abroad, but more involvement and fluid division of labor
Firm setup	Long runs, simple products	Long runs, simple products		Shorter runs, simple and more complex products
End markets	United States	United States, South Africa (Swaziland), attempts to export to South Africa (Lesotho)		South Africa
Upgrading	Process	Process, some channel, limited functional		Process, selective product
Supply chain linkages	No sourcing competence, global sourcing strategies, nominations by U.S. buyers, no local availability, local sourcing largely limited to packaging and nonstrategic services	Links to foreign-based suppliers, nominations by U.S. buyers, no reason to nominate, no local availability, local sourcing largely limited to packaging and nonstrategic services		Regional sourcing competence, no local availability, local sourcing largely limited to packaging and nonstrategic services
Skill development	Use of expatriates, language and cultural barriers to learning	Use of expatriates, language and cultural barriers to learning, general managers more embedded		Use of expatriates, skill training for complex products and shorter runs
Technology and knowledge spillovers	CMT, basic products, no local firms, no interactions	CMT, basic products, no local firms, no interactions		CMT, more complex products and production processes, no local firms, no interactions
Main constraints	TCF and AGOA phase-out, wages and productivity	TCF and AGOA phase-out, wages and productivity, availability of local suppliers and services		Skills, wages and productivity, availability of local suppliers and services

Note: AGOA = African Growth and Opportunity Act; CMT = cut-make-trim; FDI = foreign direct investment; FOB = free on board; MFA = Multi-Fibre Arrangement; TCF = Third-Country Fabric provision.

relationships with buyers. They are unlikely to nominate new textile or trim suppliers to their buyers or invest in textile or other inputs.

- *Regional South African investors*: These are increasingly important in Lesotho and Swaziland. They tend not to be directly involved in textile or other input production, but still source the majority of textile inputs from Asia, with the rest coming from South Africa. Most management positions are held by South Africans, or in some cases Mauritians or Asians. Most firms have headquarters, sales and merchandise offices, input sourcing, product development, and design teams in South Africa and run the plants in Lesotho and Swaziland as CMT operations. But some plants have more decision-making power locally. Due to the geographic proximity, there seems to be more interaction with regard to sourcing, design, and product development between head offices and manufacturing plants in several cases. These firms are not part of a global strategy but are regionally embedded, as owners have their networks in South Africa with direct relationships to South African retailers.

One of the notable gaps in the apparel landscape in SSA—outside of South Africa and Mauritius—is the lack of joint ventures (JVs) between foreign apparel firms and local investors. JVs have played an important role in facilitating the diffusion of foreign technology and know-how in the apparel sector worldwide. The experience of Sri Lanka in leveraging JVs for spillovers is illustrated in box 7.1.

Supply Chain Effects

Level and Nature of Linkages

Supply chain linkages between foreign and local firms are very limited as there are very few local firms in export-oriented apparel or input manufacturing in the three countries. In the case of Kenya, where locally owned firms do exist, there are limited interactions and linkages between foreign and local firms in terms of input provision or subcontracting. In terms of the overall value of goods and services purchased by foreign-owned apparel firms, the share from locally owned firms was less than 5 percent for all three countries (figure 7.4a). Kenya had the highest share at 4 percent of purchases. The local share in both Swaziland and Lesotho was 1 percent, and all purchases from locally owned firms were limited to services. Some foreign firms have invested in broader functions, including vertical integration into textiles (for example, Nien Hsing in Lesotho and Tex-Ray in Swaziland), embroidery, dyeing, and screen printing, but this is largely for internal consumption. There are also foreign-owned independent trim and finishing service providers, such as the YKK plant in Swaziland that supplies zippers to the region.

In **Lesotho**, there are no locally owned apparel export firms or locally owned textile, trim, or packaging companies. In addition to the foreign apparel firms, there is one Taiwanese textile mill that supplies textiles for its own apparel production and a few other apparel firms in Lesotho and

Box 7.1 The Role of Joint Ventures in Upgrading the Sri Lankan Apparel Sector

FDI and joint ventures (JVs) have played a central role in establishing and developing the apparel industry in Sri Lanka. In particular, JVs brought technology, knowledge, and skills to the largest firms in Sri Lanka that had a crucial role in the upgrading of the industry. JVs were formed between local manufacturers and the buyers of their products; or in the case of textiles and sundries, between a global supplier of the textile product and a local firm; or in three-way JVs that also included buyers. These direct relationships have permitted and furthered process, product, and most importantly functional upgrading of local firms. This is particularly the case in Sri Lanka's intimate apparel products, where (with the support of foreign investors) firms invested in equipment and human resources for specialized segments like lingerie and swimwear, and Sri Lanka became a leading global supplier.

Besides introducing modern technology, JVs also played an important role in upgrading technical and managerial skills of Sri Lankan firms. For example, Textured Jersey of the United Kingdom sent 10 managers to its affiliate, Textured Jersey Lanka, to train local counterparts over 3–5 years. Furthermore, when Textured Jersey's shares were bought by Pacific Textiles of Hong Kong SAR, China, that company based full-time seconded personnel in Sri Lanka to transfer best practices and technical expertise. The result was a significant reduction in foreign (mainly Chinese) technical personnel, in favor of locals.

Partnerships and JV relationships with buyers not only encouraged investments and upgrading in apparel production and skills, but buyers also encouraged their textile and sundry suppliers to move to Sri Lanka. For instance, due to the close relationship MAS Holdings has with Victoria's Secret, Victoria's Secret asked some of their input providers to relocate to Sri Lanka, including suppliers of various components that go into making bras and panties, like lace and pads and also warp knit fabric (Just-Style 2006).

---

the region. There are also two embroidery firms that are both Taiwanese owned and three screen printing firms, two of which are South African and one of which is locally owned (although this firm no longer supplies any foreign-owned customers). With regard to packaging, there is one corrugated paper sheet and cardboard factory and two paper and cardboard box factories that are Taiwanese owned and three plastic-related firms that produce plastic bags, plastic hangers, and plastics that are Chinese, South African and Taiwanese owned, respectively. There are also services firms that have linkages to the apparel industry, most importantly transport and freight firms. Eight transport firms were named in firm-level interviews that are largely locally owned (only two are South African). One local customs services and one local forwarder and shipping services firm were also named. For security services and business services such as accountants, IT, and legal issues, largely South African firms are used. There are also limited linkages to foreign-owned suppliers operating in the country, as only 1 percent of the total value of purchases came from this category (figure 7.4a). Overall, purchases from locally based firms are limited to packaging materials and services

Figure 7.4 Foreign-Owned Firms' Purchases of Goods and Services: Source, Value, and Type

a. Source of purchases of goods and services by value (%)

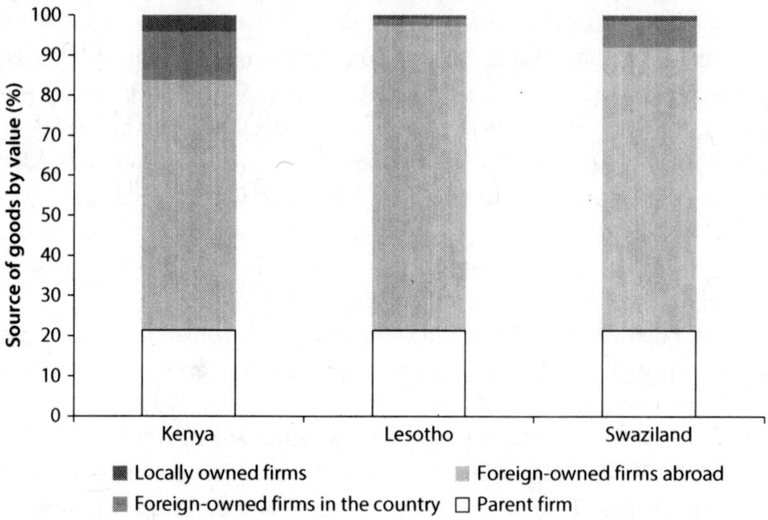

Source: Surveys of 36 foreign investors: Kenya = 15, Lesotho = 12, Swaziland = 9.

b. Purchases from locally owned and based firms by type (%)

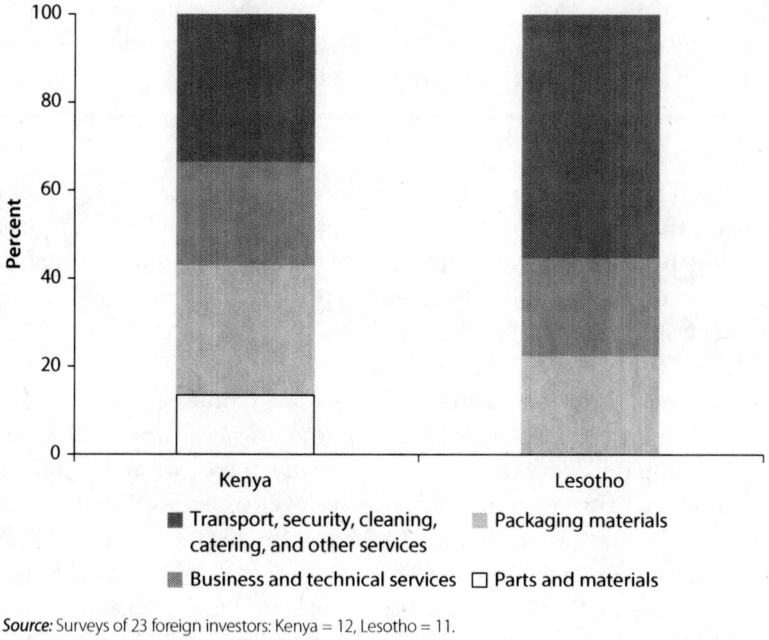

Source: Surveys of 23 foreign investors: Kenya = 12, Lesotho = 11.

Note: Swaziland is not included due to the low response rate and lack of data on the nationality of supplier firms (that is, South African or Swazi-owned). For Lesotho, there is lack of accuracy in distinguishing between South African and Basotho-owned supplier firms.

(figure 7.4b), and 97 percent of inputs by value were imported, either from parent firms or from foreign-owned firms abroad.

All export-oriented apparel firms in **Swaziland** are foreign owned with the exception of one. This firm started out in the apparel industry as a subcontractor for a few of the foreign-owned firms engaged in exporting to the United States. However, over the last decade the firm has acquired its own customers in the South African market. The firm was able to make these connections with buyers in South Africa through personal networks acquired through the original industry association set up in the country and by attending an apparel trade meeting in South Africa. The firm now maintains a split between 80 percent own contracts and 20 percent subcontracting for foreign-owned firms in Swaziland. This is a successful example of how local firms can develop through subcontracting relationships with FDI firms.

As in Lesotho, apparel firms in Swaziland are only involved in assembly and the majority of fabric and trim is imported, mostly from Asian countries. Overall, 93 percent of purchases were imported, either from parent firms or other foreign-owned firms abroad. Textile, trim, and packaging capabilities exist in the country, but they are all foreign owned, and the majority of the goods produced are for internal consumption. There is one Taiwanese-owned textile firm, Tex-Ray, with a knit fabric plant and dyeing facility that primarily provides goods for internal consumption. There is a second, foreign-owned yarn plant, Spintex, but they produce entirely for export. The only trim suppliers are also foreign owned. Tex-Ray has facilities to produce hangers, embroidery, and printing, but only for internal consumption. A Japanese-owned firm, YKK, has a zipper manufacturing facility in Swaziland, but the majority of the zippers are exported. Noncore inputs such as packaging and services come from facilities in Swaziland, most of which are South African owned. Of total purchases by foreign-owned apparel firms in Swaziland, 6 percent were from foreign-owned firms operating in the country. Only 1 percent was from Swazi-owned firms, all in the area of services such as security and cleaning.

Local apparel firms exist in Lesotho and Swaziland but they are very small scale and can be better described as individual tailors or workshops. These firms produce made-to-order products for the local markets in niche areas, including school uniforms, traditional apparel, or dresses and suits for specific events (for example, weddings). Some firms, particularly traditional apparel tailors, also export to South Africa and Botswana. This business is very distinct from apparel-exporting firms with regard to the order, design, and production process, and the equipment (machinery) and inputs used. There is no interaction between foreign-owned apparel firms and these workshops. The apparel retail sector in Lesotho and Swaziland is dominated by a few low-price retailers from South Africa, with some small, largely Asian-owned outlets. Secondhand apparel comes in from traders from Mozambique, in particular in Swaziland, and some homemade apparel comes in from the rural areas. The lack of a local retail sector further limits the potential development of local manufacturers.

In **Kenya**, there is one local export-oriented EPZ apparel firm. It primarily works as a subcontractor for foreign-owned firms, but is now moving into direct relationships with work wear and uniform clients in Europe and the United States. This firm started outside the EPZ producing for the local market and doing some subcontracting work for one of the EPZ firms. A few years later the company was able to move into the EPZ to expand its subcontracting relationship. There are still close interactions between these two firms that also involve advice in terms of production setup, productivity improvements, and quality. The firm has also recently started selling about 20 percent of output to their own clients, with the remaining 80 percent produced as a subcontractor to the other EPZ firm.

Overall, only 4 percent of the value of purchases made by foreign-owned apparel firms surveyed in Kenya was from locally owned firms. Another 12 percent came from foreign-owned suppliers operating in Kenya, and the remaining 84 percent was imported from third-party firms in other countries (63 percent) or from parent companies abroad (21 percent) (figure 7.4a). As seen in figure 7.4b, purchases from domestically owned firms were predominately services (58 percent), followed by packaging materials (29 percent) and parts and materials (13 percent).

In Kenya there are also locally owned and a few foreign-owned firms outside the EPZs, but they have very little interaction or linkages with foreign-owned EPZ firms. They are located in separate industrial areas and pursue different production and sales strategies for the local market and increasingly also regional markets. There are around 35–50 formal firms with around 50–100 machines and many more small- and micro-scale informal firms (Chemengich 2010). These firms tend to be less interested in exporting and also not able to fulfill the requirements in terms of volume, finance, and networks. For instance, formal local firms may have an output of around 300 pieces per day compared to 10–15,000 and up to 50,000 pieces per day for EPZ firms. There are also some larger textiles and apparel producers focusing on the local and regional markets that are often foreign owned. The textile sector has however contracted considerably in Kenya and there are only around 12 textile mills left that use rather outdated equipment with limited experience in supplying global markets. The remaining mills have nearly no links to EPZ firms and there seem to be no approaches to connect to them and try to supply them with textile inputs.

Interviews were also conducted with a sample of 11 supplier firms including seven packaging material suppliers and four trim suppliers (two for labels, one for hangers, and one for thread and buttons). Nine of these firms were locally owned and the remaining two were fully and partially foreign-owned. These firms primarily sell to other firms operating in Kenya—53 percent of sales were to Kenyan-owned firms in the country, 38 percent were to foreign-owned firms, and 15 percent of sales were exports.

Survey results indicate that only half of purchasing agreements between these input suppliers and foreign-owned firms operating in Kenya are formal contracts; the other 45 percent were set up as trial contracts, ad hoc purchases, or regular orders with no formal contract. Nonformal contracts create a more difficult operating environment for suppliers because they do not provide a way to plan for the future and limit the likelihood of buyers engaging in long-term relationship building, including supplier development. Out of 11 input supplier responses, six firms (54 percent) have obtained ISO 9000 quality certification; half of these firms did so as a requirement to supply a foreign-owned customer in Kenya. Only three firms stated that foreign-owned customers in Kenya provided any form of assistance to help them meet their requirements. Even when assistance was provided, the three firms perceived it as providing very minimal improvements.

The overwhelming majority of inputs are imported, given the limited existence of input suppliers (whether locally owned or foreign owned). This is particularly the case with regard to core inputs for apparel production such as fabric, yarn, trims (that is buttons, zippers, labels, threads), and machine parts. Textile inputs are almost exclusively imported. Some trims and noncore activities, including packaging and freight, forwarding, transport, and security services, are sourced locally, particularly in Kenya. These linkages are important, but due to the noncore character of these inputs and services, apparel industry-specific spillovers are limited. Table 7.6 provides an overview of the existing capabilities for each type of supply chain linkage for the three countries.

Determining Factors

Input sourcing decisions are generally made at FDI head offices, either in Asia or, in the case of Lesotho and Swaziland, in South Africa, where inputs for all production plants are sourced on a global scale. Multinational producers often own textile mills in other countries that are used to supply their apparel manufacturing plants. They also tend to source inputs on a global scale for their globally dispersed manufacturing plants to get better prices and secure conformity.

Table 7.6 Potential for Supply Chain Linkages in Kenya, Lesotho, and Swaziland

Categories and countries	Foreign-owned			Locally owned		
	Kenya	Lesotho	Swaziland	Kenya	Lesotho	Swaziland
Fabric suppliers	◐	◐	◐	◐	○	○
Apparel trim	◐	◐	◐	◐	○	○
Packaging and labels	●	●	◐	●	○	○
Capital equipment	◐	○	○	○	○	○
Subcontractors	◐	◐	◐	◐	○	◐
Local export-oriented apparel firms	n.a.	n.a.	n.a.	◐	○	◐
Yes, readily available ●		Some capabilities ◐			No ○	

Note: n.a. = not available.

A global sourcing model for textiles also limits opportunities to develop local capabilities for other less important inputs such as trim. Even if local capabilities exist for trim, if textiles are sourced from abroad, other inputs can easily be sourced abroad as well and shipped in the same box.

Another issue is the fact that many *global buyers nominate suppliers* for textiles and sometimes trim. For the three country cases, roughly three-quarters of all sourcing decisions are made by external decision makers, including parent firms abroad, buyers, or foreign sourcing agents. In the case of Swaziland, where 75 percent of supplier decisions are made outside the country, the external decision makers are primarily head offices abroad or sourcing agents, which can be explained by the larger importance of South African buyers (figure 7.5). On the other hand, in Kenya and Lesotho there is more of a split between head offices and buyers that nominate suppliers.

For firms that have sourcing competencies in the host countries and where purchases are not determined by global buyers, there is more scope for local sourcing. In Kenya, Swaziland, and Lesotho, roughly one-quarter of sourcing decisions (29 percent, 25 percent, and 14 percent, respectively) are made by local management within the country (figure 7.5). For these firms, the lack of availability of suitable local inputs and the missing link with regard to textile production is a crucial competitive concern. In Kenya, where textile and trim capabilities do exist, apparel firms perceive there to be several obstacles to sourcing from these firms. The main reasons cited were inadequate quality,

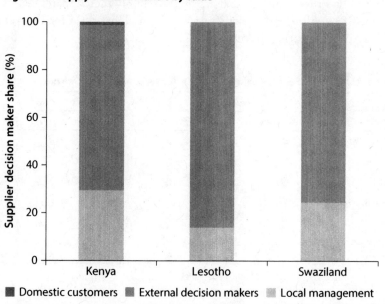

Figure 7.5 Supply Decision Makers by Value

Source: Surveys of 39 domestically owned firms: Kenya = 15, Lesotho = 13, Swaziland = 11.

uncompetitive pricing, lack of production capacity to meet scale requirements, and low levels of production technologies (see later section).

Lesotho and Swaziland have never had a substantial *textile sector* and in Kenya the sector has contracted significantly. Kenya used to have a large textiles sector (52 mills at one point) but it has contracted considerably since the 1980s and today there are only 12 textile mills remaining, with limited capacity and obsolete equipment. The Kenyan textile sector currently produces cotton and synthetic yarn, knit cotton fabric, and woven fabric; however, these products are not sourced by foreign-owned export firms, but rather are used for internal consumption (in vertically integrated firms) or as inputs by firms producing for local and regional markets.

Another domestic constraint relates to *interactions between export-oriented firms and locally owned firms*. In the case where locally owned apparel and input suppliers exist (as is largely the case in Kenya), the distinct business models, different end markets and buyers, and the different location of these firms makes interactions difficult. In Kenya, with the exception of one EPZ firm, local firms are outside of the EPZ areas. In this context, EPZ regulation is another impediment to local linkages. In Kenya, only 20 percent of total sales can be sold to customers in Kenya, Uganda, or Tanzania (given the East African Community common market) for manufacturing operations, and local sales are not permitted at all for commercial establishments (that is, importers of trim or equipment in the EPZ zone). Domestic market customers have to pay value-added tax and import duties as if the products were coming from outside the country. These EPZ restrictions would also be a significant barrier to investment in textile facilities, as the scale of these facilities would likely make them reliant on serving customers outside the EPZ.

However, the most crucial domestic constraint is the *limited existence of locally owned apparel or apparel input firms*. Even more than 10 years after the coming of AGOA and much longer after the first foreign-owned apparel firms opened in the three countries, almost no local firms have emerged to compete with the foreign firms, subcontract for them, or supply them with inputs. For Kenya in particular, with its industrial history and broader capability and skill base in the apparel and textile industry, this is surprising. The lack of local suppliers at virtually any level precludes the possibilities of spillovers through demand and technical assistance effects—including supplier requirements and assistance for standards and certification, as well as support on technical and nontechnical upgrading.

Evidence suggests there are two main factors contributing to the lack of locally owned firms in Lesotho and Swaziland. The first issue concerns government policies toward local small and medium enterprises (SMEs) investment. SMEs are subject to the same corporate tax, electricity, water, and telecommunications rates as large corporations. Funding for SME support organizations is minimal and the incentives they can offer are very limited (see box 7.2).

The second issue may be partially explained by the political economy and cultural context. First, there is little history of indigenous entrepreneurship in both countries. Second, foreign managers have noted that a reluctance to accept

Box 7.2 Subsidized Factory Shells and an Unlevel Playing Field for Local Firms in Swaziland

Local small and medium enterprise investors in Swaziland rely on the Basotho Enterprises Development Corporation for technical and financial support. However, the tools they have at their disposal are limited. The primary incentive they can offer is workspace at the Small Enterprises Development Company industrial estates. There are nine estates across the country, each with up to 20 workspaces. Rent for these spaces averages around 29 rand per square meter in the prime locations, with slightly lower rates in rural areas where there is less demand (around 24 rand per square meter). In comparison, large firms (which are almost entirely foreign owned) can access factory shells at much more heavily subsidized rates through the Swaziland Investment Promotion Agency. Based on interviews, these rates appear to be around 7.5 rand per square meter, with the first two years rent often free for large investors.

---

hierarchies between locals can make it difficult for locals to assume managerial roles. Finally, differences in business practices and styles tend to make linkages and JVs with local firms unappealing to foreign investors.

Labor Market Effects

Level and Nature of Linkages

FDI in the apparel sector has created local skills in the three countries but these skills are largely limited to basic production. In all three countries training is mostly informal, conducted by floor supervisors, and is focused on basic production and standardized assembly activities. However, there are differences between Asian- and South African–owned firms in the case of Lesotho and Swaziland, with South Africans engaging in more training, given their more flexible production setup (see following discussion; Morris, Staritz, and Barnes 2011; Staritz and Morris 2012). There is also more importance placed on skill development in more embedded global and Asian investors. This is generally not related to the more complex production process and higher-value products, as is the case with South African investors, but rather to more local decision-making power and more functions conducted locally.

To deal with the shortage of skilled labor for management, technical, and to a lesser extent supervisory positions, these positions have generally been filled by expatriates. Expatriates can have a crucial role in local skill training, but their often limited experience in engineering and management as well as language and cultural barriers result in limited knowledge transfer to local workers. There have, however, been improvements in all three countries with regard to localization, in particular at the supervisory, maintenance, and line management levels where the majority of workers are locals today. In top management there are still only foreigners in FDI firms, whereas in middle-management positions there is a mix;

foreigners tend to be in technical and financial positions while locals are in human resource positions.

In Swaziland, nearly all (83 percent) technical and management positions are filled by expatriates (figure 7.6). In Lesotho, there are few locals in the top and middle management levels, but locals do exist in areas of human resources and machine maintenance. Overall, half of the technical and management positions are filled by expatriates (46 percent). Some years ago most line supervisors were expatriates in both countries. Today, however, there is a mix of foreign line supervisors largely in charge of technical and production issues and local supervisors in charge of management and communication issues. In some firms even all supervisors are locals. In Swaziland and Lesotho, 34 percent and 19 percent of supervisors are expatriates respectively. The skill gap is smaller in Kenya, where localization has improved and the use of expatriate workers is less common than in Lesotho and Swaziland. The share of expatriates in management and technical positions in the firms surveyed in Kenya was only 32 percent and 15 percent at the supervisor level (figure 7.6). The larger number of locals in Kenya can likely be attributed to the longer history of the apparel industry in Kenya, which has produced workers with many years of experience, and the larger availability of local training institutions.

Determining Factors

As discussed, foreign firms have focused on the use of expatriates for higher skill activities while limiting training to production related tasks. However, South

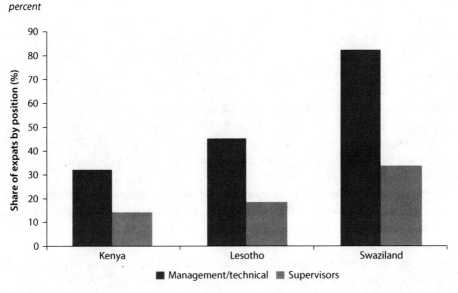

Figure 7.6 Share of Expatriate Workers in Nonproduction Positions
percent

Source: Surveys of 31 domestically owned firms: Kenya = 12, Lesotho = 10, Swaziland = 9.

African–owned firms and more embedded global and Asian investors have paid more attention to training since they produce higher-value products or more of their business functions are controlled locally. Furthermore, there are also limited industry-specific training institutes that cater to the needs of the industry. Generally in SSA, with the exception of South Africa and Mauritius, very little formal training of skilled personnel, technicians, supervisors, and managers occurs.

In **Kenya**, the Daewoo vocational training institute in the Athi River EPZ is supported by investment from Korea. There is also a textile colleague and a Department of Textile Engineering at the University in Eldorat, as well as the Kenya Textile Training Institute that is run by the Ministry of Labor and located in Nairobi. However, equipment and training is often obsolete and not aligned to industry needs. Results of the case study survey (figure 7.7a) indicate that 73 percent of respondents stated that relevant training centers existed, but only 36 percent of firms used these facilities. Firms that did not use the facilities felt they could provide better training in their own facilities, with the primary reasons being obsolete machinery and the lack of relevant focus. Furthermore, many firms stated there is a readily available pool of labor due to the diminishing size of the overall industry in Kenya. In Kenya, labor turnover was not seen as a major problem, with only 20 percent stating it as a major concern (figure 7.7b). Compared to Lesotho and Swaziland, where the large majority of local managers and supervisors have been promoted from within the organization (89 percent and 93 percent respectively), in Kenya this share is only 47 percent. This can be explained not only by the longer existence of industry-specific training facilities but also by the history of the industry in the country and the fact that there are more firms, so more opportunities for worker mobility.

In **Lesotho** there are two Skill Development Centers that were established in 2008 with World Bank support. The objective of these centers is to help the apparel sector enhance its competitiveness by improving productivity levels, adding more value to products, and diversifying into the South African and EU markets.[7] Eighty-five percent of firms reported using training facilities in Lesotho. All of the firms that did not use the centers stated that facilities do not offer training in the skills they require. Even among firms using the facilities, there were complaints that training did not cover more advanced technical and managerial skills that firms required. There is, however, more interest and engagement in the skill development centers by South African firms than by Taiwanese firms, which tend to have well-established internal training procedures. The Taiwanese firms are largely interested in basic skill training for their long-run standardized production (Staritz and Morris 2012). Labor turnover was also a problem cited by the majority of firms. In most cases, it hinders firms' willingness to train workers. The high incidence of HIV also affects firms' investments in training and workers' attitudes toward upward mobility.

In **Swaziland,** where nearly all nonproduction-related positions are filled by expatriates, a major deterrent to knowledge and skill spillovers is the inability of managers and supervisors to communicate effectively with line workers, and, to a lesser extent, lack of strategic interest in increasing local skills. There are

Sector Case Study: Apparel

Figure 7.7 Training Facilities and the Impact of Labor Turnover on Training

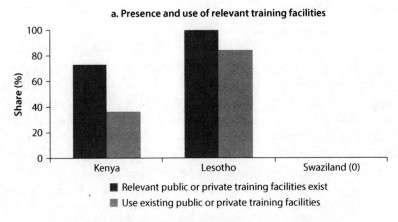

a. Presence and use of relevant training facilities

■ Relevant public or private training facilities exist
■ Use existing public or private training facilities

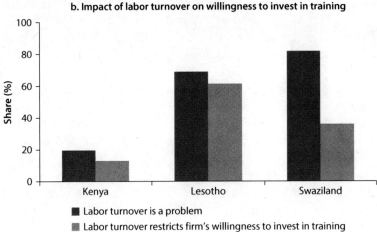

b. Impact of labor turnover on willingness to invest in training

■ Labor turnover is a problem
■ Labor turnover restricts firm's willingness to invest in training

Source: Surveys of 39 domestically owned firms: Kenya = 15, Lesotho = 13, Swaziland = 11.

virtually no apparel training centers in Swaziland. Existing machinists were originally trained on the line as are new machinists. The University of Swaziland offers a bachelor of science degree in textiles, apparel design, and management through the Department of Agriculture; however only 11 students have graduated from the program since its inception in 2003. In 11 firm interviews, all firms confirmed there was not a relevant public or private training facility, and many saw this as an impediment to future growth (figure 7.7a). In Swaziland, labor turnover is also a barrier to firms' interest in investing in training. Over 80 percent of firms viewed labor turnover as a problem, and 36 percent admitted that this deterred them from investing in training workers (figure 7.7b). As in Lesotho, the high incidence of HIV also affects firms' investments in training and workers' attitudes toward upward mobility.

Technology and Knowledge Spillovers: Demonstration and Collaboration Effects

Level and Nature of Linkages

FDI may transfer proprietary technology and knowledge related to both production and nonproduction-related activities from head offices to affiliates. The extent of this transfer depends in part on the role the specific affiliate has in the production network. As firms in the three countries largely engage in CMT production using standard machines and production technology, the spillover potential is limited from the outset. Moreover, few foreign-owned firms have undertaken significant process innovations after their initial investment, limiting the ongoing exposure of workers to new technologies.

Again, however, there are important differences across types of investors. Plants of multinational producers are part of strictly organized production networks and tend to be locked into a particular set of assembly processes as a deliberate strategy of the head offices and their global production networks. More embedded global and Asian investors are often single plant operations with a more fluid division of labor that allows for the relocation of broader functions to their plant in SSA if capabilities exist. Both types of firms predominately export basic products to the United States using relatively standard machines and production technology. South African firms in Lesotho and Swaziland in turn focus on shorter runs and more fashionable products for the South African market that require a more flexible production setup and, as discussed, some higher worker skills.

Another important aspect of this spillover channel is not directly related to technology transfer but involves learning about and accessing export and supply networks. Foreign-owned firms, in particular multinational producers, have well-established sales and sourcing networks and channels to access buyers and input suppliers. The presence of FDI firms might make it easier for local firms to access such networks. Access is even more facilitated when important buyers or suppliers establish offices in the host country. This has been the case with the international zipper manufacturer YKK and its regional establishment in Swaziland, from which it supplies apparel exporters in the region, as well as with Nien Hsing's investment in a denim fabric mill in Lesotho. Local firms could benefit from this presence as it is closer than a plant in Japan; Korea; or Taiwan, China.

Determining Factors

Again, the limited existence of local manufacturing firms that could absorb potential spillovers through demonstration is the main constraint for this spillover channel. In Kenya, where locally owned apparel firms exist, we asked these firms how they viewed their operations and how they feel they compare to FDI firms in the sector. Figure 7.8 shows how locally owned apparel producers see the degree of sophistication of their production process and the technology gap between their firm and their top foreign-owned competitor. On average,

Figure 7.8 Perceived Sophistication of Production and Technology of Kenyan-Owned Apparel Firms

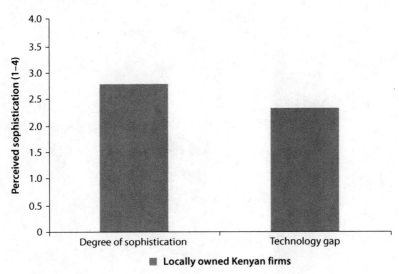

Source: Survey questions: Kenya = 13.
Note: Survey questions were as follows:
(1) On a scale from 1 to 4, where 1 means "standardized" and 4 means "highly sophisticated," how do you rate the degree of sophistication of your firm's production process today? (2) On a scale from 1 to 4, where 1 means "no difference" and 4 means "large difference," how do you rate the technology gap between your firm and your top foreign-owned competitor's technology in Kenya?

firms see their general degree of sophistication at 2.8 (on a scale from 1 to 4); the technology gap to the top foreign-owned competitor firm is seen as 2.3. Hence, local firms also view their technology as inferior to at least the FDI technology. Although this difference provides the potential for learning from foreign-owned firms, it also may limit the interest of foreign-owned firms in establishing interactions and linkages in the first place.

The majority of foreign firms in all three countries reported collaborating with other firms in the apparel sector; however the vast majority of these interactions were with other foreign-owned firms. In Kenya, 93 percent collaborated with other apparel firms in the country, yet only 14 percent of these interactions were with locally owned firms. The reported levels of collaboration were somewhat lower in Lesotho and Swaziland (figure 7.9a), but in both cases virtually all interactions remained solely within the FDI sector (figure 7.9b).

Beyond the lack of locally owned firms, there is also limited interaction among foreign-owned firms on research and development (R&D). Just over a quarter of firms in Kenya reported engaging in any type of interaction with other firms related to R&D, one firm in Lesotho, and no firms in Swaziland. Interactions that do exist are largely limited to some subcontracting relationships, sharing machines and inputs during peak production, and lobbying and coordination activities at the level of industry associations.

Figure 7.9 Collaboration of Foreign-Owned Firms with Other Foreign or Locally Owned Firms in the Sector

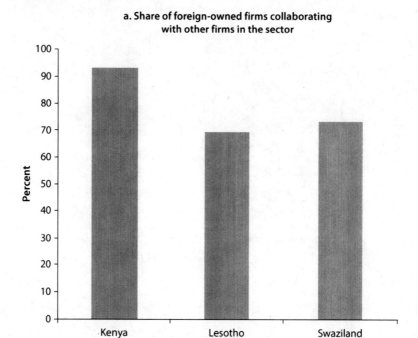

a. Share of foreign-owned firms collaborating with other firms in the sector

Source: Surveys of 39 foreign-owned firms: Kenya = 15, Lesotho = 13, Swaziland = 11.

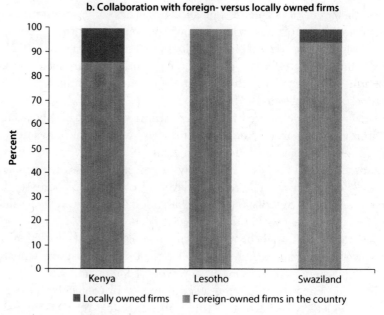

b. Collaboration with foreign- versus locally owned firms

Source: Surveys of 31 domestically and foreign-owned firms: Kenya = 14, Lesotho = 9, Swaziland = 8.

In Lesotho, these interactions between FDI are further divided between Asian-owned firms and the South African firms. These firms target different markets, operate in different geographical areas of the country, and run two independent (and almost mutually exclusive) industry associations.

Conclusions

Apparel sector FDI has benefited Kenya, Lesotho, and Swaziland significantly in terms of employment and export generation. It has also created and or revitalized operating skills and industrial capabilities and led to the improvement of trade-related infrastructure. However, the three countries have been less successful in initiating spillovers to the local economy, despite significant emphasis on attracting FDI through the use of financial incentives and instruments like EPZs. FDI has been largely related to low local value-added goods, limited local linkages and participation in management, inadequate skills development and productivity improvements, and missing local entrepreneurial response.

The limited spillovers relate to the nature of FDI and the strategic interest of foreign investors not to create such spillovers, in particular with regard to traditional multinational producers. Taiwanese firms with links to triangular manufacturing networks have a competitive advantage. Their head offices have close relationships with buyers and input suppliers, attract orders, and manage and provide higher value-added functions. At the same time, this type of network limits spillover potential in branch plants in the three countries. Beyond strategic interest, FDI spillovers are also related to local conditions and limited local skills and capabilities. Local linkages and technology and knowledge spillovers also have not developed because there has been practically no entrepreneurial response to the presence of foreign firms. Reliance on expatriates also reflects weak local technical and management skills and nonexistent or inadequate domestic training institutes. These local conditions seem to be particularly constraining for more locally embedded foreign investors, and for investors with a less well-developed and more fluid international division of labor.

As part of the broader efforts to support spillovers, policy also needs to address the fundamental challenge of sustainability of the existing sector, by improving competitiveness. This will require efforts to increase productivity and upgrade existing exporters, improve the physical and bureaucratic infrastructure in the country, diversify end markets, and resolve ongoing uncertainties surrounding preferential trade arrangements (in the case of SSA, specifically around AGOA).

With regard to policies to increase spillovers, table 7.7 lists the main factors affecting local absorption capacity and FDI spillover potential for the three channels. Table 7.8 lists the type of government support and policies that are most likely to facilitate spillovers.

Table 7.7 Factors Influencing FDI Spillovers

Spillover type	FDI absorption capacity (characteristics of host economy)	FDI spillover potential (characteristics of foreign investors)
Supply chain linkages	• Existence of local apparel manufacturers • Existence of local input suppliers • Existence of local service providers • Access to credit for investment and working capital • No segregation (physical or through regulations and taxes) of industrial areas for foreign and local investors • Minimum level of apparel manufacturers required to attract investment in textiles and (to a lesser extent) trims • For forward linkages (that is, design, marketing, retail), existence of a local market and retailers	• Specialization and outsourcing of input production (that is, no ownership of textile mills abroad) • Limited global sourcing policies • Flexible division of labor in global network • Sourcing decision power in affiliates • Supplier development programs • Foreign investors must sell to buyers that do not nominate input suppliers or open to nominating local suppliers
Labor market and human capital	• Existence of local apparel manufacturers • Availability of local supervisory, technical, and management skills • Existence of industry-specific training facilities • Good image of industry that attracts good people and motivates learning	• Provide skill development programs for workers • Expatriates that have technical and management experience and speak local language and know culture • Use of expatriates to train local employees
Technology and knowledge spillovers	• Existence of local apparel manufacturers for subcontracting or other interactions • Access to credit for investment and working capital • No segregation (physical or through regulations and taxes) of industrial areas for foreign and local investors	• Type of investment (for example, CMT activities provide limited spillover potential) • Joint ownership with local firms • Engagement in subcontracting • Participation in industry associations and other joint activities

Note: CMT = cut-make-trim; FDI = foreign direct investment.

Table 7.8 Government Support to Achieve FDI Spillovers

	Policies to improve domestic conditions	Policies to change incentives for foreign investors
Supply chain linkages	• Do not limit incentives to firms that export (also include local and regional market) • Investment fund for local manufacturers of apparel and relevant inputs and support services • Information center on requirements of foreign firms, including support for local firms • Linkage activities such as fairs bringing foreign and local firms together • Creating specific space for local firms in EPZ and other industrial areas • Initiative coordination among stakeholders along the value chain (institutions related to cotton, textile, and apparel) • Develop a regional value chain strategy	• Tax incentives for local sourcing • Local content policies in combination with local supplier development programs • Abolish restrictions on sales to local and regional markets for FDI firms • Specific incentives for FDI in textiles, for example, tax incentives or subsidized electricity costs • Approach specific foreign investors for relevant inputs (for example, textiles, complex trims)

table continues next page

Table 7.8 Government Support to Achieve FDI Spillovers *(continued)*

	Policies to improve domestic conditions	*Policies to change incentives for foreign investors*
Labor market and human capital	• Industry-specific training facilities in partnership with private sector, including components on business and entrepreneurial skills • Approach countries with successful experiences (for example, Korea, India) for training support • Improve basic education	• Incentives (for example, linked to training) for hiring locals in middle and high management positions • Training fund that supplements firm funds for specific skill training • Required cultural and language classes for expatriates at all levels • Limits for expatriates in positions where local skills exist • Support training abroad for local supervisors and managers
Technology and knowledge spillovers	• Do not limit incentives to firms that export (also include local and regional market) • Investment fund for local manufacturers of apparel and relevant inputs and support services • Creating specific space for local firms in EPZ and other industrial areas • Local sourcing for government and institutional purchases (school uniforms, military uniforms, and so forth) • Linkages activities such as fairs bringing foreign and local firms together • Initiative coordination among supporting stakeholders (government, industry associations, education and training institutions)	• Require minimum performance metrics to continue to receive incentives • Provide incentives based on the value added within the country • Incentives for foreign investors to have joint ownership with a local firm • Close loopholes that allow FDI firms to extend tax incentives indefinitely

Note: EPZ = export processsing zone; FDI = foreign direct investment.

Notes

1. The chapter builds on earlier and parallel work, in particular on joint work and regular discussions with Mike Morris from the University of Cape Town. The authors would also like to thank the representatives of apparel firms, industry associations, research institutes, and other institutions in Kenya, Lesotho, and Swaziland who took time to discuss with us dynamics and challenges in the apparel sector. Without their time and valuable insights this work would not have been possible.
2. This section is partly based on Staritz (2011).
3. GVCs can be differentiated as producer- and buyer-driven. In producer-driven chains (which are common in capital- and technology-intensive products such as automobiles, electronics, and machinery) large, integrated (often multinational) firms coordinate production networks. Control is generally embedded in the lead firm's control over production technology (Gereffi 1994).
4. Includes Bangladesh, Cambodia, China, India, Indonesia, the Lao People's Democratic Republic, Nepal, Pakistan, the Philippines, Sri Lanka, Thailand, and Vietnam.
5. There are important differences with regard to knit and woven textiles. Knitted apparel producers are more often vertically integrated into fabric and even yarn production. In the woven segment, investments in fabric weaving are substantially larger and are generally set up as independent weaving mills.

6. There were two locally owned apparel firms in Lesotho engaged in subcontracting that are no longer in operation.
7. http://www.lesothotextiles.com.

References

Appelbaum, R. 2008. "Giant Multinational Contractors in East Asia: Emergent Trends in Global Supply Chains." *Competition & Change* 12 (1): 69–87.

Bair, J. 2006. "Regional Trade and Production Blocs in a Global Industry: Towards a Comparative Framework for Research." *Environment and Planning A* 38 (12): 2233–52.

Chemengich, M. 2010. "Impact of AGOA on the Textile and Apparel Industry of Kenya." Prepared for ACTIF African Country Reports. http://www.actif.net/, Nairobi, Kenya.

De Coster, J. 2002. "Opportunities for Textiles and Clothing in Sub-Saharan Africa." *Textile Outlook International* 101 (September–October): 131–70.

Frederick, S. 2010. "Development and Application of a Value Chain Research Approach to Understand and Evaluate Internal and External Factors and Relationships Affecting Economic Competitiveness in the Textile Value Chain." PhD Dissertation, North Carolina State University, Raleigh, NC.

Gereffi, G. 1994. "The Organization of Buyer Driven Global Commodity Chains: How U.S. Retailers Shape Overseas Production Networks." In *Commodity Chains and Global Capitalism*, edited by G. Gereffi and M. Korzeniewicz, 95–122. Westport, CT: Praeger.

———. 1999. "International Trade and Industrial Upgrading in the Apparel Commodity Chain." *Journal of International Economics* 48 (1): 37–70.

Gereffi, G., and O. Memedovic. 2003. *The Global Apparel Value Chain: What Prospects for Upgrading for Developing Countries*. Sectoral Studies Series. Vienna, Austria: United Nations Industrial Development Organization (UNIDO).

Just-Style. 2006. *Victoria's Secret Gets Intimate with Sri Lanka*. October 20. http://www.just-style.com/interview/victorias-secret-gets-intimate-with-sri-lanka_id95331.aspx.

Kaplinsky, R., and M. Morris. 2008. "Do the Asian Drivers Undermine Export-Oriented Industrialization in SSA?" *World Development* 36 (2): 254–73.

McCormick, D., and H. Schmitz. 2001. *Manual for Value Chain Research on Homeworkers in the Garment Industry*. Institute for Development Studies (IDS), University of Sussex, Brighton, U.K.

Morris, M., C. Staritz, and J. Barnes. 2011. "Value Chain Dynamics, Local Embeddedness, and Upgrading in the Clothing Sectors of Lesotho and Swaziland." *International Journal of Technological Learning, Innovation and Development* 4 (1–3): 96–119.

Palpacuer, F., P. Gibbon, and L. Thomen. 2005. "New Challenges for Developing Country Suppliers in Global Clothing Chains: A Comparative European Perspective." *World Development* 33 (3): 409–30.

Staritz, C. 2011. *Making the Cut? Low-Income Countries and the Global Clothing Value Chain in a Post-Quota and Post-Crisis World*. Washington, DC: World Bank.

Staritz, C., and M. Morris. 2012. "Local Embeddedness, Upgrading and Skill Development: Global Value Chains and Foreign Direct Investment in Lesotho's Apparel Industry." Working Paper 32, ÖFSE, Vienna.

PART 4

Conclusions and Policy Implications

Main Conclusions 247
Policy Implications 263

CHAPTER 8

Main Conclusions

Abstract

Spillovers from foreign direct investment (FDI) in the short term are not necessarily positive in developing countries, due in part to competition over scarce skilled labor. Over time, however, FDI can lead to a beneficial restructuring of the entire industry, including opportunities for better-performing local participants and suppliers. But therein lies the quandary: in most developing countries, foreign investors offer the most valuable potential source of knowledge and technology to build local capabilities. The willingness and capacity of foreign-owned firms to support spillovers varies hugely across sectors and firms, and is shaped by the dynamics of the global value chains in which they operate. The scale of linkages between FDI and the local economy—particularly through supply chains—is clearly the starting point. From this comes the incentive for assistance through which knowledge and technology can be transferred. This chapter summarizes the main findings from this research presenting details on the determinants of linkages and spillovers in different value chain, country, and development contexts.

The Evidence: Does FDI Deliver Significant Spillovers in Developing Countries?

General Observations

The spillover effects of foreign direct investment (FDI) in developing countries is not necessarily positive in the short term, but can be beneficial to local participants and suppliers in the medium to long term. Evidence from the cross-country econometric exercise indicates that horizontal FDI spillovers are negative in the short term. In the presence of foreign investors, domestic firms in the same sector experience declining productivity; this is especially true for the most productive and the least productive of these firms. This finding is in line with much previous research and may reflect that positive spillovers resulting from foreign investment can be offset (partly or fully) by competition effects (that is, FDI taking market share from existing domestic firms). Perhaps most important in developing countries, foreign investors may outcompete domestic firms for high-quality labor, raising local wages for skilled labor and bidding away the best workers. However, both effects are likely to reverse in the medium to long term, once domestic firms become more productive due to increased competition and start to absorb skilled workers from multinationals.

Although foreign investors have greater inherent potential for productivity-enhancing spillovers, they are less linked with their host economies than are domestic investors. Foreign investors are generally more competitive than their host country counterparts (with higher sales, higher productivity, and greater export participation). However, we find clear evidence that they are less well linked with the domestic economy. They sell less domestically, purchase less domestically (and what they purchase is of little strategic importance), and make less use of domestic workers, especially skilled labor in managerial positions. This suggests limitations to the realization of spillover potential.

Foreign investors are also less likely to engage in the behaviors that contribute to spillovers. Further, foreign investors are much less likely than their domestic counterparts to provide assistance to their local suppliers. This matters, because technical assistance (particularly in the supply chain) is associated with local suppliers that sell more to foreign-owned firms and that are more likely to start exporting, which suggests positive spillover effects.

This does not imply that policy makers should be circumspect about foreign investment; rather, it suggests there is a case for government intervention to promote dynamic spillovers from FDI over time. Clearly, FDI delivers critical static benefits to developing countries—employment, foreign exchange, tax revenue, and so forth. Moreover, the dynamic benefits of technology and knowledge spillovers should, over time, outstrip negative competition effects in most countries. The challenge is not so much how to avoid the negative effects but how to absorb more of the positive ones. In this sense, there is clearly a need for a policy agenda to build absorptive capacity. But there is also likely to be a good case for policy to promote wider and deeper linkages on the part of foreign investors, and to ensure that local agents (firms and workers) have the necessary institutional setting to maximize spillovers.

The degree to which foreign investors engage in behaviors that contribute to spillovers depends on many factors, including FDI-specific, global value chain (GVC)–specific, and host country (firm/environment) characteristics. There appears to be cumulative causation in the engagement of foreign investors and in the realization of spillover benefits. Foreign investors tend to engage in strategic behaviors that contribute to higher spillovers where the conditions are most optimal for the productivity of local firms. This suggests there is value in intervention to support spillovers, both to encourage investors to engage in behaviors that expand and deepen linkages, and to support improved absorptive capacity of domestic actors.

Supply Chain Linkages

Local sourcing is the critical channel for delivering spillovers. Labor markets and demonstration/competition effects have important spillover potential. However, supply chains, and in particular backward linkages through local sourcing, appear to offer the most direct channel for short- and long-term gains from FDI spillovers. They also tend to be the most visible and easiest to quantify, making them particularly important for policy makers. Evidence from this study indicates that local sourcing is critical to achieving positive spillover outcomes. In the cross-country econometric results, a sector's average percentage of domestic input purchases of FDI firms in a country shows a positive and significant impact on productivity growth for low- and medium-productivity firms.

In the mining sector, one-third of all surveyed suppliers to FDI in Ghana and 42 percent in Chile started to export directly as a result of supplying foreign investors. And behavior within the supply chain matters; holding constant local purchasing, the degree to which foreign investors provide assistance to local supply chain partners affects spillover outcomes.

Both governments and foreign investors lack clear and consistently-applied definitions of "local content" and "local procurement." In many developing countries, there is strong demand and support from policy makers to promote greater use of local suppliers (or "local content"), particularly in natural resources sectors. However, uncertainty tends to exist on the part of both governments and foreign investors over the definition of "local" in at least three ways. First, does local mean sourcing from the specific communities where the foreign investors are operating, or does it mean anywhere in the country (for example, sourcing from the capital city, which may be hundreds of kilometers away and at a different scale of development)? Second, are "local" firms any firms that are registered in the country or only those owned by nationals of the country? Or even more specifically, is a distinction made between residents, passport holders, and indigenous citizens? Third, and perhaps most important, is it local ownership or local value added that matters? Most policy makers would probably agree that a foreign-owned firm employing 500 local citizens and sourcing locally to produce an input for another foreign investor delivers more "local content" than a locally owned trader who imports the same inputs and sells them on at a margin to the foreign investor. But the reality is that most situations are much more ambiguous, making broad policy related to local sourcing contentious.

All things being equal, foreign investors prefer to use local suppliers. A clear finding from the surveys conducted for this study is that foreign investors are at worst agnostic about using local suppliers. All things being equal (which of course, they are not), FDI would much prefer not have to rely on importing goods and services, and they would benefit from having suppliers with whom they can interact on a face-to-face basis and that can respond quickly when needed. This is perhaps obvious, but is an important starting point from a policy perspective.

However, incentives used to attract export-oriented foreign investors in the first place can create a bias against local integration. This may happen, for example, if a duty is waived on imports but value-added tax or other taxes are still incurred for domestic purchases. Restrictions on the flow of goods and people inside and outside a special economic zone (SEZ) or administrative barriers for trading between the zones and the domestic economy can also create disincentives for foreign investors to engage with local suppliers.

In contrast to the limitations of the apparel and mining sectors, agribusiness may hold much more promise, looking forward. The evidence suggests that the apparel sector is so burdened with trade-and-rules-of-origin constraints as to be an outlier for potential backward linkages. Despite 20–30 years of FDI in several case study countries, virtually no domestic linkages have been established in this sector. And large, modern mining operations are so capital-intensive—with great economies of scale and need for sophisticated engineering equipment and processes—that establishing supply chain linkages beyond local services runs up against significant limitations. In contrast, the agribusiness sector is more inherently reliant on local goods and services inputs, and effective spillovers in the agribusiness supply chain offer significant scope to promote exports and support inclusive growth. Moreover, the sector can benefit from the wide range of models and experiences on linkages that already exist around the globe.

The experience of supply chain linkages is generally very poor in low-income countries, but evidence suggests it is possible to build meaningful linkages over time. Evidence from the surveys and case studies indicates very low levels of purchasing of goods and services from local suppliers in developing countries, particularly in African countries and the apparel sector. However, Ghana's experience in the mining sector (to take one example) shows that it is possible to develop some local presence in foreign supply chains over time, by establishing the right conditions and market incentives, and building on existing local capacity.

Global supply chain management trends reduce the opportunities for local supply participation. Across all value chains assessed in the study, there is a clear trend toward global supply chain management, which tends to result in the most strategic and high-value purchases being coordinated on a global or at least regional level. This has the potential to create significant opportunities in countries that may be regional source markets. However, for most low-income countries, it will establish significant limits on spillovers through domestic supply linkages.

The best short-term opportunities in low-income countries come through outsourcing of noncore services. But there is a tradeoff, as these activities are less likely to deliver dynamic spillovers. Not only are most supply linkages in low-income countries very limited in scale, but they also tend to be concentrated almost exclusively in low value–added, noncore services activities (with some nonservices activities like packaging). For example, in Lesotho and Swaziland, the most common activity that was considered a locally provided input was security services; beyond these come cleaning and basic maintenance and, further up the value-added ladder, catering. The noncore nature of these inputs matters for spillovers; this is because the provision of assistance to local suppliers tends to be much more likely when the goods and services they provide are core parts of the *upstream* value chain. For example, in the agribusiness sectors, local firms that provide core inputs to FDI in agriprocessing are most likely to receive assistance; in the apparel sector, cut-make-trim (CMT) subcontractors are most likely to receive assistance.

Assistance effects in supply chains tend to be relatively narrow in their focus, but the emphasis on quality and standards still represents a significant area of potential for upgrading of domestic firms. Not surprisingly, investors tend to concentrate their assistance efforts on issues that are related to their specific needs. For example, financial support will focus on meeting short-term working capital (to avoid delays in production and delivery) but not on longer-term finance (which would enable suppliers to carry out investments to improve productivity and embed spillover benefits). In terms of activities, this means support tends to be linked to FDIs' own compliance requirements—such as health, safety, and environmental (HSE) and quality issues. On the other hand, even when quality and standards are firm specific, they are often built on global foundations and have the potential to play a significant role in upgrading the capacities of local suppliers, enabling them to serve other investors or to start exporting.

Labor Market Linkages

Foreign investors make relatively greater use of local skilled staff than they do of local suppliers in developing countries, but this varies significantly across countries. It is not surprising that, on the whole, foreign investors rely substantially on local labor in their operations. But in the highly skilled technical and managerial positions, where some of the most important knowledge spillovers are likely to take place, use of local staff is far from certain. Overall, foreign investors do make significant use of local workers even in skilled and managerial positions, but this varies across sectors and countries. In the mining sector, for example, 70–80 percent of workers in skilled positions in Chile are local, while across African countries surveyed, the share ranges from 30 to 50 percent. In agribusiness, 75–85 percent of management, supervisory, and technical workers in Kenya and Vietnam were local, while the figures were 10–15 percentage points lower in Ghana and Mozambique. Finally, in apparel, while more than two-thirds of management and technical staff are local in Kenya, less than 20 percent are local in Swaziland.

Foreign investors prioritize local technical and managerial staff, up to a point. As is the case with supply chain linkages, foreign investors generally prefer to use local staff in most skilled positions. Survey results indicate that by far the biggest constraint perceived by foreign investors to hiring more local staff in technical and managerial positions was lack of skilled labor. There are, however, some caveats to this finding. First, most foreign investors continue to reserve certain positions for foreign nationals for reasons of corporate culture or to ensure effective communications with the head office. Second, language and cultural factors matter; survey evidence suggests there is a greater use of foreign workers where there is a significant language gap between the host country and the foreign investors. Finally, the cost of supporting foreign workers also plays a role.

Foreign investors with access to high-quality workers from their home markets at relatively low wage and relocation costs are less likely to feel the financial pressure to hire local managers.

In developing countries, spillover benefits through labor markets are significantly constrained by limited labor mobility and entrepreneurship. As noted previously, foreign investors often outcompete domestic firms for access to a limited pool of highly skilled workers. This is confirmed by evidence from our surveys, which show that firm characteristics differ markedly between foreign and domestic firms in almost all respects, with the notable exception of worker skill profiles. Employment in foreign-owned firms tends to offer significant advantages over domestically owned firms, including higher pay and benefits, opportunities for career advancement, international mobility, and prestige. This can act as a barrier to skilled labor circulation between foreign and domestic firms. As a result, diffusion of knowledge tends to be largely restricted within the FDI sector. This obviously varies by sector, and appears to be a bigger issue in apparel and mining (in most countries) and somewhat less so in agribusiness. The situation is aggravated by relatively low levels of entrepreneurialism in many low-income countries, restricting the potential for diffusion through firm spin-offs. Indeed, the advantages of FDI employment may create a further disincentive to engage in entrepreneurial activity.

Training offers an important channel for knowledge diffusion, yet this too is constrained by labor market factors and by an emphasis on firm-specific rather than transferrable skills. Despite the advantages that foreign investors hold over domestic firms in labor markets, labor turnover (normally involving workers moving between foreign-owned firms) is cited as a barrier preventing greater investment in training and skills development of the workforce. This is most evident in agribusiness, to a lesser degree in apparel, and much less so in mining. While FDI do invest moderate levels in training, this tends to be focused on company-specific skills, which may limit transferability. Partly for this reason, foreign investors tend to make limited use of local training facilities, again restricting opportunities for knowledge diffusion. Again, the use of local training facilities appears to be greater in the agricultural sector than in mining or apparel.

Spillovers from Competition, Demonstration, and Collaboration

Competition effects may result in negative rather than positive short-term spillovers in low-income countries, although this may be due to the fact that positive spillovers take more time to materialize. As noted previously, findings indicate negative short-term productivity effects (that is negative spillovers) at the horizontal level from FDI in developing countries. This suggests that negative competition effects (domestic firms losing market share and thus increasing per unit costs, along with labor market competition) are outstripping the positive productivity spillovers from competition and demonstration effects. A large part of the explanation is probably temporal—that is, positive competition and demonstration effects take time to emerge whereas the impact of negative competition effects can be observed more quickly. But it may also be that in low-income countries, lack of absorptive capacity restricts the potential for countries to benefit from positive competition and demonstration effects.

Demonstration effects are most prominent in tightly organized supply chains where the local supply base is large and fragmented. Foreign investors have an incentive to promote demonstration where providing individual technical assistance is prohibitive and/or inefficient. This is most apparent in the agribusiness value chain, where foreign investors actively promote demonstration effects to support upgrading of their suppliers through the establishment of demonstration plots or nucleus farms.

But spillovers from demonstration are constrained by generally low levels of collaboration between foreign investors and domestic firms in the same sector. Findings from the case studies and surveys indicate that in most countries sector collaboration is weak, particularly between foreign-owned and domestic firms. Most collaboration is contained within the foreign investor community (or, separately, with the domestic sector) and for purposes unrelated to knowledge and research (for example, lobbying government, addressing common labor issues, and so forth). Of the three sectors studied, only agribusiness showed any significant levels of collaboration between foreign firms and the domestic sector, particularly through links with national training centers and research institutes.

While setting standards is important, assistance appears to be critical to supporting spillovers. Evidence from the surveys suggests that demand effects alone—for example, carrying out technical audits or requiring that local suppliers make specific changes to products or processes—may have limited impact on spillovers in low-income countries. Instead, the provision of technical assistance, with or without corresponding requirements of suppliers, contributed to greater spillovers. This suggests that while the proliferation of global standards within GVCs may create an opportunity for firm upgrading, most firms in developing countries will require active support in order to take advantage of the opportunity.

Domestic firms tend to benefit more from knowledge transfer through investments in foreign machinery and equipment than directly through their links with foreign investors. Case study and survey evidence indicates that local suppliers in developing countries tend to benefit from knowledge and technology embedded in the capital equipment they purchase, and from the technical assistance they receive from equipment vendors. This may be linked to supplying a foreign investor, but it need not be so. This highlights the fact that foreign investors are far from being the only source of knowledge spillovers. It also underscores the importance of open trade policies that ensure domestic producers can access the highest-quality goods, services, and equipment from global markets.

The Determinants: What Mediating Factors Shape the Nature and Extent of FDI Spillovers?

General Observations

Having a stable, quality investment climate is a necessary starting condition to benefit from FDI spillovers. Spillovers obviously are possible only when a country is successful in attracting FDI in the first place. In this regard, our surveys indicate that three factors stand out (in order of importance): (a) political and social stability, (b) business environment, and (c) access to land and facilities. In addition, for resource-seeking investment, raw materials are critical. That the two most important factors driving FDI are so fundamental and broad highlights the fact that the policy agenda for spillovers cannot be narrow and operational but must step back to address broader investment-climate issues. The third factor (access to land and facilities) points to the potential importance of infrastructure, including industrial parks and SEZs; these have the potential to support spillovers but can also erect barriers that block them.

In sectors that are strongly linked to global production networks, trade policy considerations also play a critical role in determining FDI. For export platform investment, particularly in the context of GVCs, investors are looking for locations that allow for efficient importing of inputs and exporting of components or final products. The most attractive environments for this will have favorable locations and effective trade facilitation, but they will also be influenced significantly by the trade policy environment. Of the sectors analyzed in this study, the apparel sector is the most affected in this regard. Investor location determinants in apparel have long been driven by trade policy considerations. These started with quota-hopping investment during the Multi-Fibre Arrangement (MFA) era, and continue today as investors seek locations that allow for duty-free access to key markets while maintaining flexibility on sourcing of inputs under the U.S. African Growth and Opportunity Act (AGOA), the European Union's Everything But Arms, and so forth. Even outside tightly integrated production networks, trade policy matters for attracting investment. For example, mining investors require a policy environment that allows them relatively easy importation of capital equipment and technical services. This suggests the importance of an open trade policy environment and is therefore a caution against using trade policy as a tool through which to promote local sourcing.

Generally, absorptive capacity and host country characteristics matter most for spillovers, but FDI characteristics also matter in that they establish the potential for spillovers. As will be discussed in detail later in this chapter, the findings of this study point to absorptive capacity and host country characteristics (which also shape absorptive capacity) as the most important factors determining the scale and nature of spillovers from FDI. This finding is nothing new—a long line of research has come to this same conclusion. However, the study also highlights an important role for FDI characteristics, particularly in the context of GVCs. FDI helps determining what potential exists for spillovers in the first place, in defining their scope and nature, and in shaping the diffusion process.

In many developing countries, a large share of the supplies, services, and skills demanded by foreign firms simply does not exist. The most common response given by foreign investors concerning their lack of domestic sourcing is that required products or services do not exist in domestic markets. This may reflect an information failure, in which case a possible intervention is supporting greater access to information on local suppliers. If domestic suppliers cannot meet quality requirements, another possible intervention is upgrading suppliers' capacity. But in many small developing countries, the reality is that no local suppliers exist for a share of the large-value, strategic inputs required by foreign investors.

In natural resources–intensive sectors, the peripheral location of investments places serious constraints on linkage and spillover potential. In some sectors like mining and agriculture, production often takes place in remote parts of the country. These areas are distant from the economic center, sparsely populated, economically disadvantaged, and poorly connected to domestic and international markets. This creates a double challenge in the context of spillovers. Where investors might find it challenging to source suppliers and skills in the country overall, finding them in remote locations will be substantially more difficult. At the same time, political pressure will be even stronger for investors to deliver spillovers that specifically affect remote areas.

The presence of a large or specialized neighboring country can be a significant barrier to generating supply linkages in some sectors. Foreign investors would prefer where possible to purchase goods and services inputs from local markets. However, some items (such as capital goods, highly specialized goods, and services inputs) can only be supplied economically from regional locations (if not global). Here, only a few countries will emerge as winners. Typically these will be the larger and more advanced developing countries (for example, in the African context, countries like South Africa, Ghana, and Kenya), leaving smaller, developing countries with a narrower set of opportunities. In some situations, this effect can extend to much less strategic products and services. For example, in the case studies of Lesotho and Swaziland (in apparel) and Mozambique (in mining), foreign investors made extensive use of South Africa for virtually all their input requirements, due to the relative proximity of the market and the tight market integration that exists.

Language and culture appear to play a significant role in the process of establishing linkages and generating spillovers. While the sample is small and much of the evidence anecdotal (although also supported by some survey findings), the study suggests that the choices of suppliers and managers and the way in which investors integrate (or not) with local markets is influenced by language and cultural factors. Not surprisingly, investors have established international supply relationships that tend to be linked to their home countries and communication is obviously critical for shaping both supply chain and labor market spillovers. Yet policy tends to remain blind to this.

FDI Characteristics and Practices

Joint ventures between foreign and local firms deliver greater spillovers than either foreign or domestic firms alone. Results from the analyses conducted in the study suggest that foreign ownership has a positive impact on spillovers; but joint venture arrangements are even more positive. The surveys show that partially foreign-owned firms are more likely to source and to sell domestically, and the econometric results confirm higher productivity spillovers from partially foreign-owned firms. But the best results come in joint venture arrangements, when foreign firms have control. This may be because foreign investors have a greater incentive (and less risk) to share knowledge under this arrangement.

Market-seeking FDI is more likely than efficiency- and resource-seeking FDI to deliver spillovers. Econometric and survey results confirm that market-seeking foreign investors have greater linkages and more positive spillover effects than either resource- or efficiency-seeking FDI. Market-seeking foreign investors are, almost by definition, more forward integrated and would thus tend to have greater need for local management. But they also are more likely to source from local markets and to provide assistance to suppliers than efficiency- or resource-seeking investors (the latter do the least local sourcing). Efficiency-seeking FDI appears to be the least integrated and delivers the least spillovers. This is partly because their investment decisions favor locations with low wages with market access. These tend also to be locations with gaps in supply capacity and in the absorptive capacity of suppliers and workers. This is most apparent in the apparel sector case studies.

The time horizons of investors matter—activities, sectors, and firms that are investing for the long term are more focused on developing domestic linkages. Short-term investors are less likely to invest in integrating with the local economy than investors planning to be operating in the host country indefinitely. This is confirmed by econometric results showing a positive association between spillovers and a multinational's presence in the country. For investors with a short time horizon, the cost of recruiting and training local staff and of building the capacity of local suppliers is unlikely to pay off. From a sector perspective, this may also help explain why apparel sector investors (which have limited capital investments and operate on seasonal contracts, with tight margins and shifting sources of competitive advantage) have less incentive to invest in spillovers. By contrast, agribusiness and mining investors tend to have much longer time horizons. Even within these sectors, however, the situation can vary dramatically from one type of mine/crop to another.

Investors in sectors with higher rent potential are more likely to engage in activities to support spillovers. Related to the finding on time horizons of investors, the study also finds that profitability matters for spillovers. In sectors like apparel where profit margins tend to be tight and firm success and even survival is far from guaranteed from year to year, investors simply have less time and resources to put toward building local capacity. Sectors with large rent potential, most notably mining, face a very different situation—they can afford to invest in activities to promote spillovers and indeed face much greater pressure to do so. From a policy perspective, this has implications in terms of what is reasonable to expect from different types of foreign investors.

The use of ad hoc rather than formal, long-term contracting significantly restricts the likelihood of spillovers. A striking finding from the study was the relationship between contract types and the likelihood of foreign investors providing assistance to local suppliers. In the agricultural sector, for example, while more than a quarter of firms on formal contract received assistance from foreign investors, this share dropped to only 16 percent for firms operating on ad hoc contracts. And for most of the African countries covered in the survey, ad hoc contracting seems to dominate. In Ghana and Mozambique, only around 20 percent of suppliers in the agriculture and mining sectors indicated they had formal contracts of greater than one year in length. By contrast, in the Kenyan agribusiness sector, the share was 43 percent, in the Chilean mining sector it was 48 percent, and in the Vietnamese agricultural sector it was 83 percent. One possible reason for the greater use of ad hoc contracting in Sub-Saharan Africa is that the nature of the products and services being bought locally does not lend to formal, long-term contracting. In any case, the implications appear to be significant, as foreign investors have less incentive to invest in the supply relationships, reducing significantly the potential for spillovers. Another spillover determinant is the transparency of contract laws. Poorly designed, missing, or nontransparent contract laws may act as a barrier to foreign investment, and also disadvantage local firms in competing for supply contracts with foreign investors.

Source country characteristics also matter—regional investors tend to be most integrated with domestic markets, other emerging-market investors the least integrated, and investors from mature markets in the middle. Results from the survey analysis indicate that investors from Sub-Saharan Africa have built much deeper linkages to domestic markets in African host countries than their counterparts from other regions. This is true in terms of sourcing local content, selling to local markets, and employing local technical and managerial staff. While African investors showed a strong positive association with linkages as well as assistance to local suppliers, investors from other emerging markets showed just the opposite. They were much less likely than the average foreign investor to buy or sell in local markets, make use of skilled local staff, or provide assistance to their local suppliers. This contrast is notable in the apparel sector, where recent regional investment from South Africa and Mauritius tends to operate a very different model from the traditional AGOA-focused foreign investors (from Asia and elsewhere). In addition, the survey and case study evidence suggests that investors coming from mature markets that face pressures for "sustainable investment" (including demands for attention to quality, HSE, labor standards, and more generally contributions to host country development) are more likely to engage in activities to deepen linkages.

The sophistication and scale of domestic firms not only determines spillovers through supply chains but also through labor markets. A similar story holds for spillovers through labor markets. A number of countries have been successful in generating spillovers through labor markets. For example, in the apparel sector, success stories in Bangladesh, Sri Lanka, and Mauritius are linked to the transfer of skills, knowledge, and networks from foreign investors to local skilled workers and entrepreneurs. There are similar stories, on a smaller scale, in the Ghanaian mining sector and the Kenyan and Ghanaian agribusiness sectors. But again, this requires having skilled workers in managerial and technical positions in the first place and having local firms to which these individuals could move (or an environment that fosters entrepreneurial ventures in the sector). In many low-income countries, particularly smaller countries, these basic conditions are not in place.

Better education and skills of workers contribute to greater supply chain spillovers, but only for larger and well-run firms. Bringing together the last two points, the study finds that education matters for spillovers. However, the econometric exercise indicates that higher education and skills improve spillovers for medium- and high-productivity firms, but not for low-productivity firms. This suggests the interaction between workers and managers (or to put it another way, the combination of labor productivity and technical efficiency) is critical to absorbing spillovers.

Host Country Characteristics
In Sub-Saharan Africa, many of the barriers that constrain competitiveness of domestic firms are the same ones that block greater linkages from being achieved. Insufficient trade and transport infrastructure, low-quality and high-cost utilities, regulatory barriers, and lack of access to affordable finance, among other barriers, raise the cost bases and reduce the quality of output of domestic firms. Combined with weak skills and management capacity (as discussed previously), these factors also tend to keep local firms in low-income countries small. They find it difficult to achieve the scale and technical capacity to supply foreign investors at a cost and quality that is competitive with alternative sources. It also places severe restrictions on the potential of integrating forward and backward in value chains. For example, the high cost of electricity virtually closes off opportunities for extending apparel investments backward into fabric and yarn. While electricity accounts for some 5 percent of input costs in apparel, this rises to 20 percent in weaving and knitting and to 35 percent in spinning. Similarly, high energy costs make adding value to agricultural and mining commodities unprofitable.

Labor market mobility makes only a limited contribution to spillovers in low-income countries. Results from the econometric analysis indicate than open, flexible labor markets have a positive impact on spillovers from partially owned foreign firms, but not fully owned foreign firms. This is likely to be partly a function of the relative labor intensity of partially versus fully owned foreign firms. However, it also supports previous findings that higher-end skill shortages in low-income countries may be a significant constraint to the diffusion of knowledge between foreign and domestic firms.

Government spending on education favorably influences FDI spillovers. Econometric results indicate that a country's government spending on education as a percentage of GDP has a strongly positive and significant productivity effect, confirming the positive role of the local innovation infrastructure for FDI spillovers.

University-industry linkages can play an important role in supporting labor market spillovers. Results from the case studies indicate that university-industry linkages are relatively weak across most sectors and most countries. However, they also suggest that the presence of strong linkages would be valuable in supporting spillovers. For example, foreign investors in the agribusiness sector in both Kenya and Ghana have played a role in addressing specific skills gaps in the sector through curriculum development for universities and putting in place industry placement programs. These efforts have contributed to the spillover of knowledge to new entrants in the labor market, many of whom would go on to work in domestic firms or establish their own firms and farms.

Trade and financial market openness is critical for facilitating spillovers. Results from the econometric exercise indicate a positive impact of both trade and financial market openness on spillovers, especially from partially foreign-owned investors. One obvious reason for this is that FDI is more likely to be present in the first case with a higher degree of trade and investment openness. However, the findings also suggest that openness contributes to higher spillovers even in the presence of FDI. FDI may facilitate the absorptive capacity of domestic firms, possibly because they are able to operate more competitively with access to quality inputs and finance for investment in technologies and skills that support absorption. Indeed, the evidence suggests the positive impact of financial market openness is strongest for high-productivity firms.

Lack of a strong legal framework for contract enforcement can be a barrier to developing supply linkages in many countries. Foreign investors in both the agribusiness and mining sectors identified contract enforcement as a significant barrier to developing linkages with local suppliers. This came out as a particular issue in Sub-Saharan Africa and with small and medium enterprise suppliers. In the agribusiness sector, the concern was primarily around the difficulty of enforcing contracts with smallholders. In mining, the limited ability to take legal action against suppliers in case of nondelivery or delay in delivery of mine-critical inputs made it unmanageable to source many higher value-added inputs from the local market.

The Role of Global Value Chains

For low-income countries, GVCs offer expanded possibilities to attract FDI, but actually raise greater barriers to embedding and extracting value (spillovers) from it. There is increasing interest from policy makers in developing countries to "join" or "integrate into" GVCs. The presence of GVCs opens up opportunities for developing countries to attract significant volumes of FDI and quickly establish a presence in new (and often diversified) sectors. It also links developing countries in global networks governed by standards and technologies that have valuable potential for upgrading of domestic firms and workers. However, realizing this spillover potential is another issue. Indeed, the barriers to spillovers may be even higher than they are in non-GVC environments. This is because GVCs—with global governance of supply chains and often footloose investing—create an environment where FDI may have less incentive to invest in deepening supply, research, and labor market integration in host countries, and where technologies and processes for production may have been significantly disconnected from local realities. The implication is that the process of upgrading within GVCs may be severely curtailed. This risks the sustainability of investment in the first place, as locational attractiveness remains reliant on access to inputs (labor or natural resources) whose price cannot remain suppressed indefinitely. This scenario is apparent in the apparel GVC investment in Lesotho and Swaziland, for example. Investment came quickly, but after more than 20 years, the FDI sector remains almost wholly disengaged from the domestic economy and no upgrading of the sector has taken place. The result is a situation where uncertainties over trade preferences and wage pressures constantly threaten the viability of investments.

Spillover potential varies significantly depending on the prevailing governance structures in specific GVCs. Different structures of value chains—for example buyer- versus producer-driven chains—are likely to offer different possibilities for low-income countries to benefit from spillovers. For example, the apparel and agribusiness sectors (particularly agribusiness linked to the global retail sector) are buyer-driven GVCs, where many of the supply chain decisions, technologies, and standards are driven down through the top of value chains. The result is that the foreign investor operating in the developing country may actually have little to no influence on the extent and nature of linkages (at least in terms of supply linkages). This is because the ultimate buyer may well prescribe exactly what inputs they should source and even the supplier (globally) from whom they should source it. In apparel, for example, the surveys indicate that local purchasers only make decisions on 29 percent, 25 percent, and 14 percent of purchases in Kenya, Swaziland, and Lesotho respectively.

Buyers also establish the standards that must be met, which may deny local firms any possibility of participating in the supply chain. For example, a European supermarket could impose requirements that suppliers must meet EurepGAP standards for all ingredients going into a product. In this case, a foreign investor producing that product in Ghana may have to source certain inputs from outside the domestic sector if it is not in a position to meet EurepGAP standards. For example, domestic sector firms may not be able to provide traceability from farm to factory.

Value chain dynamics vary significantly even within sectors. Even within sectors, GVC dynamics may differ considerably, raising or reducing barriers to linkages. For example, different mining commodities have very different technology requirements, which may facilitate or preclude domestic sourcing. They also have very different development time spans, which as discussed previously affects the incentives for FDI to invest in building local relationships. A similar situation holds for the agribusiness sector, where certain commodities are more capital and technology intensive that others and time frames for development and exploitation of crops varies widely. Even the stage of activity may determine spillover possibilities. In mining, for example, exploration offers limited supply and labor market spillover possibilities, whereas construction offers substantial direct employment and sourcing potential but relatively limited knowledge diffusion (and a very limited timeframe). Mine operation offers the best possibilities for broad spillovers in supply chains, labor markets, and competition/demonstration.

Across all GVCs, increasing supply chain consolidation and tiering raises barriers to spillover potential for low-income countries. Despite the significant variation across GVC types, there is a common trend toward consolidation of supply chains, including management on global and regional levels. Associated with this is the increasing use of tiered supply chain structures: lead firms in GVCs outsource entire components or stages of activities to "tier one" firms, who are then responsible for sourcing and production of subcomponents within that stage. This tiering structure is most well known in automotive and electronics, but it is also apparent in the apparel sector (with the emergence of "full package suppliers") and in mining, where the large mining houses routinely hire engineering, procurement, and construction management (EPCM) firms for the construction phase, and in some cases even outsource mine operations to third-party firms.

This has important implications for spillovers in developing countries. First, local firms in low-income countries will be increasingly likely to operate in the second or third tier of supply structures, with effects on value addition and, therefore, on margin possibilities (see following discussion). Second, these firms will be increasingly less likely to have direct interaction with the global lead firms, which may constrain access to new technologies and knowledge. Finally, related to this, the actual foreign investors entering developing countries in many sectors will often not be global lead firms but rather their tier one or tier two suppliers. As discussed earlier, these firms may have less incentive and be less capable of delivering spillovers.

But value chains are dynamic, and changes will continue to create opportunities as well as barriers to achieving spillovers. Many of the value chain dynamics discussed have emerged or transformed only in recent years. Changes in technologies in particular will continue to affect the opportunities for local supply and human capital linkages, as well as research and other network collaborations. The implications of these changes may be positive or negative for local supply linkages. For example, in the mining sector the development of new reprocessing technologies has enabled local firms in South Africa and Chile to establish profitable reprocessing operations in the gold and copper industries. Other technology improvements in the sector have encouraged greater cooperation between mining companies and their suppliers to improve operational efficiency. This has created opportunities for increased spillovers to local suppliers, but in some cases it has established barriers that prevent local suppliers from participating. In the apparel sector, improvements in fabric and production technology can also act as a barrier to linkages (for example, through subcontracting) in low-income countries, as firms lack the capacity to invest in required new equipment.

Value chain positioning and the subsequent impact on the profit margins of local suppliers may be a significant barrier to absorbing spillovers and upgrading. Because of developing countries' low position in value chains, local firms tend to have squeezed profit margins. Thus, they will typically not be in the financial position to invest in the upgrading of technology or skills that would be required for absorbing spillovers. For example, a local cocoa processor in Ghana identified from FDI processors a set of process improvements to raise productivity (demonstration effects). But value chain dynamics forced the firm to sell through a local broker rather than directly to multinationals, significantly eroding the firm's margins and preventing it from implementing the upgrading investments.

GVCs are driving the growing importance of global standards, with ambiguous implications for spillovers in low-income countries. As discussed previously, the emergence of global standards is one of the most important implications of GVCs. Global standards present a significant opportunity by which knowledge and technology from FDI can be transferred in a codified way to firms and workers in developing countries. At the same time, these standards raise barriers to local firms, as compliance requires skills, knowledge, and investment. Moreover, demands for compliance to strict standards by lead firms (especially buyers) in value chains may result in more risk-averse behavior, such as restricting relationships to established suppliers from established markets.

In some cases, the imposition of standards within GVCs has had a positive spillover effect to the domestic regulatory environment. This has been seen in the adoption of HSE standards and, in the case of Kenya, in the adoption of agricultural and food standards (for example, KenyaGAP).

With South-South investment within GVCs growing rapidly in importance, the implications for spillover potential in low-income countries is mixed. Across all three GVCs studied, the share of foreign investment coming from developing and emerging markets is large and growing rapidly. As discussed previously, evidence on their implications for linkages and for spillovers is mixed. On the one hand, South-South investment may lead to greater absorption through the use of more "appropriate" technologies. On the other hand, evidence that emerging market investors may set a lower bar on standards and quality may ultimately have detrimental effects on the competitiveness of local suppliers. Moreover, the model of GVC dynamics that structure FDI—domestic economy relationships across different sectors—is built primarily on the experience of the operations of Western multinationals. Much less is known about what patterns will emerge from the governance of South-South value chains.

CHAPTER 9

Policy Implications

Abstract

The challenge of realizing positive spillovers from foreign direct investment (FDI) in developing countries is huge. Policy makers should be aware of the difficulty in achieving local economy linkages from FDI, much less spillovers, and recognize that many factors outside of their control will determine from the outset the scale of these opportunities and their achievability. But government policy and programs can make a difference. Government has a role to play as a provider of information, as a facilitator, and as a regulator. In this chapter we summarize the policy implications that emerge from the findings of this research, providing details on the role that government—along with foreign investors, the domestic private sector, and other stakeholders—can play in promoting deeper FDI integration into domestic economies, and in harnessing more effectively the potential of FDI spillovers for development. In sum, the generation of backward linkages and local supply chains depends on creating a favorable investment climate for local firms no less than for foreign investors, including: access to finance and imported inputs, enforcement of contracts, reliable regulatory standards, adequate power and other infrastructure support, and adequate competition in the domestic economy. These are the necessary, although not sufficient, conditions for success. Added to these cross-cutting factors, light-handed but deliberate and well-targeted programs that work to support building capacity and competitiveness in domestic firms can increase the likelihood of positive outcomes.

Introduction

This final chapter outlines a set of policy implications resulting from the findings of this research. These are organized across three sequential areas of policy (figure 9.1): (a) attracting the "right" (most strategic) investors in order to maximize the spillover potential; (b) promoting effective linkages between foreign investors and the local economy, and ensuring that investors have the incentive to engage in the activities and behaviors that should contribute to realizing spillovers; and (c) establishing an environment where local actors (firms, workers,

Figure 9.1 Policy Framework for Spillovers

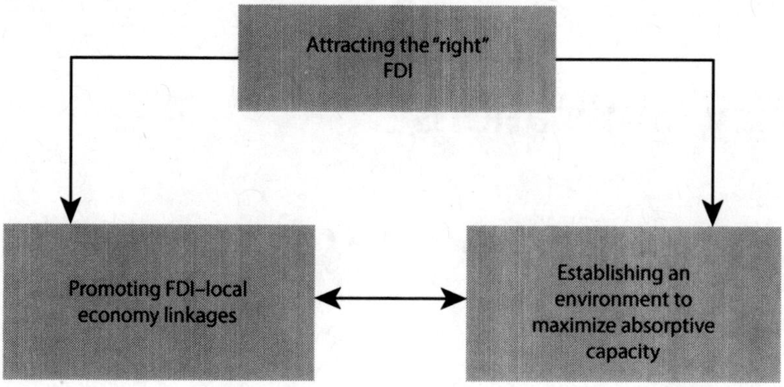

```
                    ┌─────────────────────┐
                    │ Attracting the "right"│
                    │         FDI          │
                    └──────────┬──────────┘
                   ┌───────────┴───────────┐
                   ▼                       ▼
        ┌──────────────────┐      ┌──────────────────┐
        │ Promoting FDI–local│◄────►│ Establishing an  │
        │  economy linkages  │      │ environment to   │
        │                    │      │ maximize absorptive│
        │                    │      │    capacity      │
        └──────────────────┘      └──────────────────┘
```

Note: FDI = foreign direct investment.

and entrepreneurs) have the capacity to take advantage fully of this spillover potential. In the final subsection, we discuss the institutional and implementation arrangements for putting in place these policies.

Before delving into the specific policy issues it is worth first asking two more fundamental questions about spillovers policy. First, *how much priority should governments give to actively facilitating spillovers?* Second, *to what degree must policy interventions be sector (or global value chain [GVC]) specific?*

Prioritizing the Extent of Support for Spillovers

It is not a foregone conclusion that government should spend substantial resources trying to promote spillovers, despite the benefits that may accrue from them. One of the main reasons is that, as discussed throughout this book, the dynamics of GVCs may limit the influence of government policy. Particularly for small, low-income countries and in nonresource-seeking value chains, governments have relatively limited power to either attract the "right" investors or to ensure those investors are integrated deeply into their domestic economies. One of the important messages of this book is that governments need to be realistic about the linkages and the spillovers that are realizable, particularly over the short term. That being said, government can play a role in ensuring that local actors are in a good position to absorb whatever knowledge and technology diffusion may result from foreign direct investment (FDI) presence. Moreover, market failures do exist—particularly around information as well as investments in training and standards—which justify government intervention. Finally, where governments have the power to influence FDI behavior (for example, in sectors where rents are significant), evidence suggests that policy incentives to promote greater integration of FDI into local supply chains and labor markets can have an impact.

Cross-Cutting versus Value Chain–Specific Interventions

This conclusion suggests that both are relevant. The most significant (and resource intensive) interventions that governments can make to influence spillover possibilities should be targeted at supply-side absorptive capacity. For the most part, these interventions (in skills, access to finance, and so forth) are cross-cutting in nature. On the other hand, some aspects of capacity-building investments are likely to have sector-specific components. Two examples of this interplay between cross-cutting and sector-specific interventions come in the areas of quality and vocational training. Improving absorptive capacity of local firms and workers may require investments to improve the national quality and vocational training infrastructure and institutions. But they may also require specific investments in order to enable firms to meet specific sector quality standards and to deliver sector-specific technical skills. Moreover, interventions aimed at overcoming information failures—for example the typical linkage program activities like supplier databases and buyer-supplier matching—are necessarily sector specific in nature.

Attracting the "Right" Foreign Investors

The starting point to generating spillovers is attracting FDI in the first place. One of the key messages of this book is that not all foreign investors are the same when it comes to their potential to deliver spillovers. Therefore, given the increasing priority being placed on generating spillovers from FDI, governments will need to take into account the optimization of spillovers more explicitly when developing investment promotion strategies and policies.

Countries tend to focus on the active promotion of a sector without first putting in place their investment strategy for development. It would therefore be important to make a clear distinction between the functions of investment policy coordination and those of investment promotion. Investment promotion is typically carried out by an investment promotion agency or intermediary. Policy coordination should be placed at a ministerial or prime minister level with convening power over and full coordination of the agencies in order to ensure long-term linkages/spillovers policies.

Following is a summary of the main policy messages from this book relating to investment attraction.

The most important policies to attract strategic GVC-oriented FDI for spillovers should remain focused on ensuring an attractive general investment climate and policy environment conducive to trade. GVC-oriented investors are hardly different from other investors in their main requirements for determining investment location decisions: political and macroeconomic stability and the general business environment matter most. From the perspective of investment promotion strategy, this means the policy advocacy role—promoting investment climate reforms—should remain among their highest priorities. In this context, trade policy and especially the trade facilitation environment is the highest priority for GVC-oriented investors (particularly export-platform investment). Thus, investment promotion agencies may need to put particular emphasis in promoting trade policy reforms.

Investment policy, promotion strategies, and efforts to promote linkages should account explicitly for the nature of investment and the motivations of potential FDI. The objectives of investment promotion will of course go well beyond realizing spillovers. But to the degree that spillovers are part of the objectives, strategies need to recognize that the potential for establishing FDI–local economy linkages and for realizing spillovers, at least in the short term, are likely to be very different if the investment is of the efficiency-seeking/export platform or the resource-seeking variety versus market-seeking. Similarly, sectors and specific multinational corporations (MNCs) will vary in their potential to deliver spillovers. To the degree that spillovers are a high priority, governments should target investors with the following characteristics: market-seeking (at least partly); operating in sectors that rely on localized inputs; export to mature markets (if efficiency seeking); integrate but do not internalize value chains; and those with experience in establishing and developing local suppliers and talent. In addition, regional and diaspora investors can be seen as valuable, priority targets. An important caveat here is that for many countries, particularly smaller developing countries, the realistic range of possibilities for attracting investment may be rather narrow. But these optimal targets can be built into identification and targeting strategies as well as incentives policies, including access to fiscal incentives and to privileged land and facilities (for example, in special economic zones [SEZs]).

For large-scale investments, particularly of a resource-seeking nature: (a) target top-tier MNCs with a track record in supporting local economy linkages; and (b) make preparation of a local economy spillover strategy part of the package for consideration in awarding exploration and development licenses. The potential to target specific investors and to feed spillover objectives directly into the investment attraction process is obviously much greater for large-scale foreign investment projects that are likely to deliver substantial long-term rents. This is most common in the mining sector, but other resource-seeking investments (in some agricultural sectors) and some services sector investments may also qualify (for example, in energy, construction, transport, property, and potentially financial services). In cases where potential investors are, in effect, competing for access to a concession, an exploration license, or some other exclusive right, submissions on their strategies and commitments related to promoting linkages and spillovers could be made one component of the evaluation.

Assess (appropriate) technology contribution as an explicit element of the FDI evaluation process. It goes without saying that developing countries should have a clear process by which potential foreign investment is evaluated, particularly in the context of provision of incentives. Part of that process should be an evaluation of the potential technology contribution of the investment, which could include a focus on the degree to which the technologies that investors may bring are likely to be absorbed in the economy, given current capacity.

Target promotion efforts beyond original equipment manufacturers (OEMs) and lead firms to tier one global suppliers. The changing structure of GVC investment makes it increasingly important for investment promotion efforts to widen beyond the traditional, well-known lead firms to focus on their tier-one suppliers. In doing so, it is important to also recognize that many of these investors will have less experience and capacity in engaging in active programs to support spillovers. It will also be important to ensure that both requirements and incentives to promote spillovers should be pushed down below the lead firms to include the investors to whom they contract out operations (this has been a common problem in the mining sector). Once some tier one firms are established, options should open up to attract tier two firms (with guaranteed contracts to tier one firms). These firms are even less likely to have the capacity to engage in active supplier development, but they are more likely to invest through joint ventures (JVs) and work closely with local partners.

Avoid bidding away the benefits of spillovers by excessive incentives to attract FDI. One important finding of this research is that realizing positive spillovers from FDI is far from certain, more so for low-income countries and potentially more so from GVC-oriented investment. Yet governments routinely use spillover potential as a justification for offering subsidies and other incentives to foreign investors. It is noteworthy that incentives tend to be most commonly associated with attracting export-platform investment, given its more footloose nature. Yet the findings of this research suggest that it may be particularly challenging to realize spillover benefits from this type of investment. Thus, governments should be particularly wary of using substantial fiscal incentives to attract GVC-oriented investment.

Recognize that the "right" investment to deliver spillovers requires both foreign and domestic investors—ensure that investment policies do not bias against domestic investors and that they support mutual interaction. Investment policies need also to be supportive of domestic investors, including both competitors of FDI as well as potential partners and suppliers. This means that domestic investors should not be unfairly disadvantaged by the incentive regime, and they should have access to infrastructure and services that are designed to support the establishment of globally competitive firms. One example of this is SEZs, which are often established primarily for the benefit of foreign investors and may have explicit or de facto barriers to domestic investors. Opening up such instruments to domestic investors not only helps promote their productivity and level the playing field, but can facilitate greater FDI integration through physical proximity and networks.

Facilitate JVs where it can be value adding, but avoid coercion. JVs appear to be an effective channel for facilitating spillovers, particularly of older technologies and know-how (which, for low-income countries, are likely to be most relevant). This finding, however, should not be misread to argue for attempting to force investors to engage in JVs with local partners. First, it is important to recognize that the correlation depends a bit on context. Market-seeking investors are far more likely to seek JVs, and they are also far more likely to have deeper linkages with local suppliers and workers. For export-oriented investment, however, most multinationals will have a strong preference for 100 percent ownership. Second, there is obviously a big difference between demand-led JVs and forced partnerships—open information sharing on the part of the multinational is far less likely in the latter case.

While regulating JVs or technology licensing is not recommended, governments can help provide information to potential investors about the qualified domestic investors that could be viable partners. This information provision role could also be effective in helping local investors identify international partners, as well as subsidizing local firms to gain exposure to international networks.

Encourage investment in value-added processing (beneficiation)—where viable—as a source of potential backward linkages. Particularly in natural resource–intensive sectors, there is a tendency to argue that adding value to commodities is the main way in which foreign investment can contribute to economic development. There are, of course, strong and well-established arguments that in most developing countries and for most commodities, few investments in value-adding activities are likely to be successful and sustainable. The results of this research suggest that value-added processing is particularly effective for establishing economy-wide linkages. In this context, governments should promote FDI in value addition, being careful to avoid subsidizing projects that are not likely to be viable.

Use industrial policy in a light-handed way that focuses mostly on overcoming market failures or capturing coordination externalities. Weaknesses in institutions, in private sector capacity and organization, and in skills and absorptive capacity are the norm in low-income countries. This context raises an array of challenges to fostering linkages, making it difficult for policy makers to prioritize interventions and strike the right balance with limited available public funds. Some countries (for example, Chile) have been successful with policies promoting a favorable investment climate across all types of industries and investors. However, many developing countries will feel the need to target particular sectors that are seen as promising in terms of their linkages and welfare effects. The trick is to fashion a light-handed industrial policy that focuses mostly on overcoming market failures or capturing coordination externalities, including packages of infrastructure expenditures and public-private vocational training initiatives. But in promoting linkages through targeted sector strategies, it is important that those chosen sectors conform to reasonable projections of comparative advantage.

Promoting FDI–Local Economy Linkages

Having brought foreign investors into the country, the next set of policy considerations are around how to ensure that investors are as integrated as possible into the domestic economy. The logic here is that strong linkages with the domestic economy—through supply chains, labor markets, and other forms of collaboration and interaction—should result in greater diffusion of knowledge, technology, and know-how from foreign investors. Note that the most important actors here are the multinationals. Government's responsibility is to establish the right incentives to encourage the integration of foreign investors into the domestic economy. Broadly speaking, government can do this through the stick of regulation, the carrot of incentives, as well as through other actions like facilitating information flows. All three have some role to play. In this context, following is an overview of the main policy messages emerging from this study.

Ensure the incentives used to attract investors in the first place do not create a bias against local integration. Incentives available to export-oriented foreign investors typically include import tax and duty concessions or duty drawbacks. Such instruments play an important role in ensuring competitiveness by giving access to international quality inputs at competitive prices. And for inputs like capital equipment and sophisticated technologies, import markets are in any case the only source. Governments should, however, ensure that such instruments are aligned to remove disincentives they may create for developing local supply bases. Most importantly here, the issue is to ensure that foreign-owned companies do not have privileged access to such instruments over domestic producers. Similarly, as noted previously, reserving SEZ access to foreign-owned companies can create barriers to supply by domestic firms. This may happen if a duty is waived on imports but value-added or other taxes are still incurred for domestic purchases. But there may be more subtle barriers created, for example by restricting the flow of goods and people inside and outside the SEZ gates, or by creating administrative barriers for trading between the zones and the domestic economy.

Leverage investment incentives to promote actions that support spillovers. If the realization of spillovers is among the principal rationales for offering incentives to foreign investors, then those incentives should be predicated if not on spillover outcomes (difficult to measure) then at least on investors engaging in activities to support spillovers: these include local supplier development, provision of technical assistance, training of workers, joint research, and so forth. For example, in Australia's mining sector, the implementation of an Industry Participation Plan is required to access tariff concessions offered to investors in the sector.

Local content regulations have a role—under the right conditions and if tightly circumscribed. Establishing a clear policy framework, including regulations aimed at promoting linkages—and even specifically "local content"—can play a constructive role in promoting linkages.[1] Case study evidence suggests that the introduction of legislation or even just discussions over its possibility send the message that government is serious about addressing FDI linkages, and can therefore encourage a coordinated response from industry. In terms of policy and regulation specifically aimed at expanding local procurement, a number of options may be implemented. These include set-asides and targets for local sourcing, and procurement processes that give preference to local suppliers and workers.[2]

But regulations can only be effective when the domestic supply side is actually up to the task of being a competitive supplier to industry. Introducing local content requirements before local industry can respond adequately is likely to weaken the competitiveness of investors, undermining the overall objectives. In any case, setting strict local content targets can be counterproductive and difficult to enforce.

Tools and measurement systems, along with capacity building, should be put in place to ensure proper ex ante evaluation of policies that promote spillovers. In the absence of such tools, governments risk developing ad hoc regulations that are poorly suited or can even work against the objectives of establishing FDI spillovers. Many countries have implemented local content regulations that establish specific quantitative targets for localization, without first understanding the current level of procurement of local goods and services, or what would be a realistic target given the scale and capacity of the domestic private sector. Greater research is needed to understand what kind of tools would be most effective for governments to employ for such ex-ante evaluation.

Instead of rigid local content requirements, focus on collaborative development of flexible localization plans. A more effective approach to facilitating FDI–local economy linkages is for government to encourage (or oblige) investors to come up with their own proposals on how they will deliver spillovers to the local economy, allowing for flexibility so that different sectors and firms contribute substantially to improving linkages in ways that are efficient and sustainable. As discussed previously, this can be done as a requirement for obtaining a concession, an operating or development license (particularly in the case of natural resource sectors), or to access investment incentives or other instruments. Ghana's local procurement policy in the mining sector is an example of good practice; the government has managed to strike a balance between pushing obligations on firms and providing some flexibility as to how those obligations are met. Such plans can be expanded beyond the narrow confines of local procurement to include other activities critical to facilitating spillovers, such as the provision of technical and financial support to firms, training and skills development, and local firm and institutional collaboration.

Define "local content" clearly and focus on value addition. In establishing any incentives or regulations to promote local economy linkages, and specifically to promote localization or "local content," governments must set out clearly—for agreement by all stakeholders—what is meant by these terms. Is local referring to anything in the country or does it have a specific geographical reference to areas affected most directly by the foreign investment? Are local firms those that are fully owned by locals, have a majority of locals on the board, or just registered in the country? Is a local a resident of the country, a citizen of the country, or only a certain citizen (for example, an indigenous citizen)? Even more important is the definition of "content." Here, governments should focus on promoting in-country value addition rather than simply in-country ownership. A developing country is likely to benefit a lot more (both in static terms but also, critically, in spillover potential) from a foreign investor manufacturing locally and employing

Policy Implications

locals than it is from a local-owned firm that imports virtually everything and sells on with a margin. Figure 9.2 presents a continuum for defining local firms, based on levels of local participation and the extent of value added in the country.

Bridge information gaps by facilitating exchange of information on investor needs and local supplier capacity, as well as skills requirements. One of the clearest roles for government in supporting improved domestic market linkages is to address information gaps. Government can ensure foreign investors are aware of potential domestic suppliers and their capabilities, while also giving suppliers access to information on the specific goods and services that FDI are seeking, along with requirements in terms of quality, certifications, scale, delivery expectations, and so forth. This information can be delivered by undertaking vendor registration, establishing databases, and facilitating exchange through exhibitions and active supplier matching. Beyond pure information provision, efforts can be made to facilitate the development of supplier networks and sectoral learning networks. Similar activities can improve information availability for specific skills requirements, as well as for matching FDI research and technology requirements with the programs and capabilities of local universities and research institutes. Such efforts are most often delivered by government, working through the investment promotion agency. However, while existing information failures motivates government facilitation and a funding role, it is not necessary for government to actually carry out the work. Sector bodies, nongovernmental

Figure 9.2 Framework for Measuring Local Content

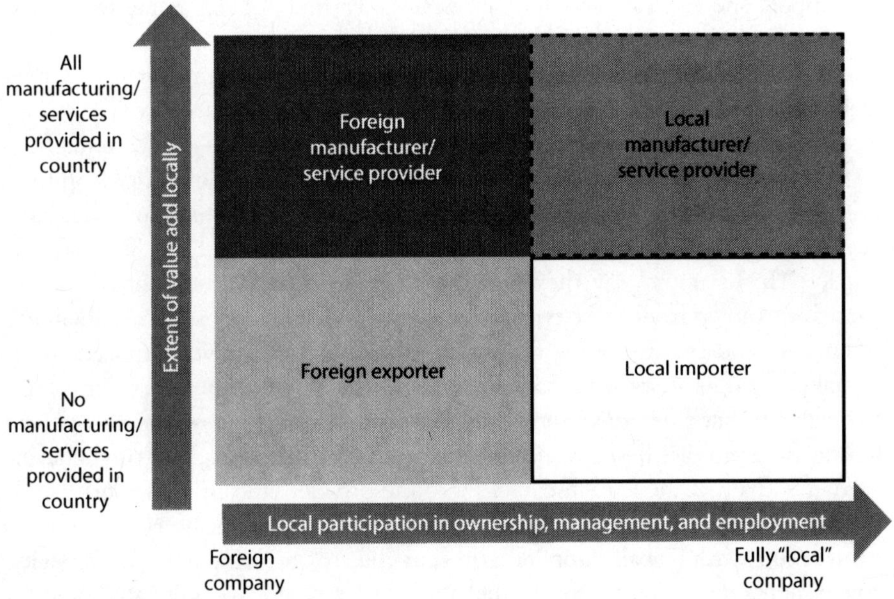

	Foreign manufacturer/ service provider	Local manufacturer/ service provider
	Foreign exporter	Local importer

Y-axis: Extent of value add locally (from "No manufacturing/services provided in country" to "All manufacturing/services provided in country")

X-axis: Local participation in ownership, management, and employment (from "Foreign company" to "Fully 'local' company")

Source: World Bank 2012.

organizations (NGOs), and specialized private service providers have experience, networks, and capacity to lead these efforts.

Be realistic about what type of local content is achievable in the short term but calibrate supply side support programs where the most value added is possible. Governments in low-income countries need to be realistic about the level and nature of linkages that will be established in the short term, in terms of supply, skilled labor, and industry collaboration. In many cases, supply opportunities will be limited to basic services activities like security, maintenance, and catering. Similarly technical and managerial employment opportunities may be limited mainly to supervisory type positions. This "low hanging fruit" should be targeted immediately. Resources to support supply side capacity building, however, should be focused on medium-term objectives, to firms and workers that can be involved in higher value–added activities. MNCs should also be encouraged to focus on upgrading potential—the objective is not simply creating the linkage but on providing upgrading assistance through comprehensive supplier development programs.

Having a clear and comprehensive framework for supporting the upgrading of domestic firms is important for facilitating supplier development programs that will be initiated by foreign investors. MNCs clearly need to take a lead role in planning and implementing supplier development programs. But they will be much more effective in doing so if they are able to link their program into a wider framework established and supported by government. This will enable them not only to access additional funding but more importantly to leverage complementary activities. For example, operational assistance to supplier firms can be implemented in parallel with more theoretical management and technical training that may be provided through government support programs; technical support and consultancy on quality can be carried out in parallel with assistance on certification from the national quality bodies; and technical advice might be linked with suppliers taking advantage of a program to invest in new technology.

Traditional linkages programs merely scratch the surface—they are likely to be effective only in the context of a more comprehensive set of policies on linkages. A corollary to the above point is that traditional linkage programs that have been championed by donors and some large MNCs are likely to have limited impact. This is not to say they have been unsuccessful—a number of success stories are known from these types of programs. One concern is the sustainability of linkages and the upgrading process in the absence of active support to individual companies. A broader concern is the issue of scale. These programs tend to be able to handle a maximum of 20–30 firms in each 1–2 year cycle. Specific linkage programs are best led and financed by the private sector, but can be leveraged effectively if they are integrated into the broader linkage framework.

Address gaps in domestic contract enforcement and other barriers to formal contracting with local suppliers. In parallel to building supply capacity, governments also need to ensure that the regulatory environment that governs supply relationships supports the establishment of deep linkages. A crucial regulatory goal is to facilitate more long term, formal contractual relationships, which create much greater incentives and opportunities to facilitate knowledge and technology spillovers. One important component of this is contract enforcement—weak enforcement is a serious constraint to foreign investors entering into contracts with local firms, particularly small and medium enterprises (SMEs). Another component is the transparency of contract laws—poorly designed, missing, or nontransparent contract laws may act as a barrier to foreign investment, and also disadvantage local firms in competing for supply contracts with foreign investors. Improving national credit information systems is another avenue for supporting formal contracting. Finally, intellectual property law (and its enforcement) remains an important barrier to knowledge and technology sharing from FDI.

Establish incentives for foreign investors to engage in collaboration with local universities, research institutes, and training institutes. While most attention tends to be focused on supporting local procurement and localization of skilled staff, an important channel for potential spillovers is the collaboration of foreign investors with local institutions. Incentives can be made available to promote collaboration on research and to support technology licensing and technology transfer. These include the creation of research funds, matching grant programs, or fiscal incentives (for example, tax deductions) for carrying out research and development activities in the host country. Similarly, incentives can promote collaborative activities on skills development with universities and vocational training institutions; these include supporting internships, outplacements, and joint training and curriculum development.

Encourage foreign investors to use innovative, flexible approaches to procurement. A number of well-established practices exist to integrate the potential of local firms (especially SMEs) into FDI supply chains. These include breaking procurement into smaller lots, establishing parameters for contracting with groups of small firms, offering accelerated payment terms, and upfront payment.

Establishing an Environment That Maximizes the Absorption Potential of Local Actors

Attracting investors and integrating them into the domestic economy should create optimal conditions for local firms and workers to benefit from spillovers of knowledge and technology. But the degree to which they ultimately benefit depends crucially on the absorptive capacity of domestic actors (which in turn influences the degree of domestic economy integration foreign investors will seek, as well as the attractiveness of the host economy in the first place). This is the area of spillovers policy where government has the most important role to play, but also where some of the most important policies and programs that can support spillovers are the least specific and targeted. While there is a broad range of interventions that government can take to build absorptive capacity, the policy recommendations set out here are built around a framework in which government has two roles: *(a) building the absorptive capacity of firms and workers*, and *(b) helping local firms and workers to access opportunities.*

Facilitate an incentive regime that is aligned with the development priorities of low-income countries. In low-income countries, FDI—especially through JVs—may be the only and the most effective way of upgrading the local supply (quantity and quality), due in part to limited domestic investment. FDI, as well as nonequity forms of foreign investment, such as franchising or contract management schemes between foreign companies and domestic investors, may also help address the problem of certification and standards, areas in which local SMEs are often struggling. Low-income countries are thus in a particular need of favorable investment policies that will make the entry and sustaining of investment less risky and costly, while also more attractive. FDI in services may be particularly important in this regard, as this kind of FDI tends to generate employment in higher value–added jobs, enabling transfer of know-how to the domestic human capital.

Programs to support supply-side capacity building need to take into account the heterogeneity of domestic firms. The potential to supply foreign investors and to upgrade in higher value–added activities varies enormously across domestic firms. Supplying foreign investors should be seen as an activity for the most-productive, high-potential domestic firms (supplying FDI can be seen as a proxy for exporting from the perspective of firm sophistication and productivity), and government programs focused on upgrading technical capacity should focus primarily on these high-potential firms and set out clear requirements for firm participation. For example, Chile's "World Class Supplier" program aims to promote innovation and upgrade technology capabilities in the mining supply sector. It sets a requirement that participating local firms are "recognized in Chile and abroad," export more than 30 percent of their goods or services, and have highly innovative and technologically advanced services.

Building absorptive capacity of local firms requires both general and industry-specific investments to upgrade technical capacity and, most importantly, achieve quality standards. In addition to general support for building capacity of firms, industry-specific upgrading of technology and technical capabilities is required to absorb spillovers from FDI. Government support may be most effective in helping firms to meet international quality standards (and be certified against this). In doing so, it may be important to draw a distinction between general standards (for example, ISO) and industry-specific ones. It is the latter that may be most relevant for local firms to enter international supply chains and benefit from spillover opportunities within them.

In implementing the above, flexible delivery and financing models are required that allow for sector-specific approaches and collaboration with FDI. The sector-specific nature of some of these investments requires effective collaboration with industry bodies, which play an important role in shaping and/or delivering the necessary technical support and training. In Chile, for example, training programs to build local supplier capacity are designed and implemented in partnership with industry associations, both in agriculture and mining. Various financing models, including matching grants, vouchers, and tax deductions, give greater flexibility for delivery by nongovernment actors.

Facilitate investment financing for local suppliers by teaming up with banks and leveraging FDI supply relationships. For firms to absorb spillovers and sustain their upgrading, ongoing investments are typically required in technology, process improvements, and skills. Leaving aside the potential programs and incentives discussed previously, firms will require access to affordable finance in order to make these investments. Therefore, government efforts to promote affordable access to credit (through financial sector reform, information provision, as well as incentives) remain important to delivering spillovers. Innovative approaches here could involve leveraging existing supply relationships with FDI to improve the credit terms that banks extend to local suppliers (for example, by exploiting the assets and reputation of the lead firms as a partial guarantee on suppliers).

Provide incentives for licensing of technology from foreign investors. For relatively high-capacity local firms, access to technology from foreign investors can be a significant source of spillovers. Incentives for licensing and transfer of technology have proven to be effective in promoting upgrading in many developing countries. One caveat here is that the findings of the study stress the importance of focusing on support for the transfer of appropriate technologies.

Ensure the provision of support goes beyond the technical arena and focuses on basic business and financial management. Survey and case study evidence in this book highlights the need for supply-side support to extend beyond technical upgrading and address more fundamental aspects of business and financial management of local firms. To enable them to meet the expectations of foreign investors, local firm support should include general management and accounting training, training on procurement and record keeping, and provision of small business tools. This is particularly relevant in low-income countries. Many of the specific technical needs of FDI can be supported best through their direct involvement in supplier development programs. However, the broader skills-building efforts might be best integrated with existing government programs for SMEs.

Promote scaling-up of fragmented local suppliers in order to raise productivity and enable them to engage more effectively with FDI. Particularly in sectors like agriculture, fragmented local producers can be encouraged to establish collective and/or cooperative structures. These can help producers achieve greater scale economies, allow for investments in common goods (for example, warehousing and other facilities, marketing, distribution, testing and quality control), and pool knowledge and expertise. Cooperative upscaling can improve overall productivity and put local suppliers in a better position to engage with FDI.

Demonstration pilots can be efficient to promote spillovers of technical and process know-how. Again, in fragmented value chains like agriculture, government can invest in demonstration pilots (of new technologies, processes, standards, and so forth) as a way to diffuse knowledge and know-how coming from international markets.

Both industry-specific and general education policy is critical to achieving spillovers in the long term; reducing the technical and managerial skills gap with FDI should be a priority. Establishing a domestic environment with the capacity to absorb knowledge and upgrade on a continuous basis depends more than anything else on the quality of its human capital. Building this capacity is necessarily a long-run project and, therefore, less attractive politically and out of sync with the shorter-term nature of most linkage and spillover programs. But education and skill building should be clearly established as the most important component of any spillovers program, and specific targets should be set for reducing the technical and managerial skills gap with FDI.

Engage universities and research institutes actively to embed spillovers. Governments can promote collaboration with local universities and institutes (on both research and skills development) through a range of incentives discussed previously. Beyond this, government can support greater collaboration by investing in high-quality public research, through technically advanced and effective industry extension programs (for example, in agriculture), and by supporting global standards (though a strong and efficient regulatory environment and an effective quality infrastructure).

Open policies to promote imports and skilled immigration may actually be critical to promote localization in the long term. A policy of openness, not only on access to imported goods and services, but, more controversially, on access to (imported) skilled workers, is likely to pay off in the long run, by improving the sophistication and competitiveness of local firms.

Institutional and Implementation Arrangements

Finally, this section provides brief recommendations on issues to be considered in implementing spillovers policy, including targeting, institutional arrangements, coordination, and monitoring.

Spillovers policies should be integrated directly into national industrial policies. Efforts to generate spillovers should be seen and managed as part of wider policies to promote the development and competitiveness of domestic industry (including goods and services). As noted previously, many of the policies and programs to support spillovers, particularly supply-side capacity building, clearly extend beyond the narrow realm of investment policy. In order to bring in the necessary expertise for planning and delivery, and to ensure effective coordination of efforts across government ministries and agencies, spillovers policy (like FDI policy more broadly) needs to be integrated with industrial policies and SME development policies.

Many developing-country governments will need to build capacity in their own institutions in order to implement spillovers policy effectively. Delivering the necessary supply-side support to promote spillovers requires expertise and capacity that many, if not most, developing countries will not have developed fully. An important part of the process of implementing spillovers policy will be upgrading domestic delivery capacity, particularly in terms of the national quality infrastructure, technology transfer offices, and vocational training.

Responsibility for delivering the spillovers agenda should be taken at a senior ministerial level rather than as an add-on activity for the investment promotion agency. In line with the importance of integrating spillovers as part of wider industrial development policy, the responsibility for development and implementation of policies to support linkages and spillovers should rest at a senior ministerial level, led by a strong champion in government. It should not simply be another responsibility hoisted on the investment promotion agency, which has increasingly been the case in many developing countries over recent years.

Given the huge potential base of beneficiaries in the domestic supply sector, targeting will be necessary. One of the fundamental challenges to promoting spillovers is that building supply-side capacity runs into the same scale issues that plague SME policy. With limited resources, how is it possible to avoid interventions simply being a drop in the ocean? Thus, targeting of recipients will be necessary. One solution is to concentrate initial efforts on more-advanced domestic firms (which in most cases include medium and larger local firms), such as those that have already established successful supply relationships with foreign investors. This would avoid wasting resources on firms that may never be successful in establishing a relationship with FDI. However, at the same time it raises the risk that the support simply crowds out private investments and/or fails to support those firms that could benefit most from it. An alternative would be to adopt a self-selection mechanism, like matching grants.

Another approach for targeting and efficient delivery may be to implement some supply-side interventions through existing industry clusters. Programs aimed at supply-side capacity building, including technical upgrading, technical and business management training, and quality certification, may be delivered at the cluster level, if there are organized, capacitated clusters. This approach offers an effective way to reach a wide range of firms in a related industry, and cluster network effects also have been found to be particularly successful in facilitating knowledge spillovers.

Where technical support is provided to domestic SMEs, include concurrent financing support. Building capacity of SMEs to supply or collaborate with foreign investors is necessary but often not sufficient to establish sustainable linkages. Support on financing, through improved access to working capital and investment finance, tends to be required in parallel with technical support. This is a common approach in many linkage and supplier development programs.

Finding sustainable funding for linkage and spillovers programs should be a priority at the outset. One of the main problems with typical linkage programs is that they are designed with a short life (often 2–3 years), due to limited availability of donor funding or unwillingness of private corporate sponsors to commit beyond a short time frame. In sectors that deliver significant tax revenues, royalties, or dividends (for example, mining), government may earmark revenue from FDI to fund linkage/spillover programs.

Matching grant programs can be another way to develop sustainable funding and can help crowd-in additional financing from foreign investors. Another way to extend available finance is to establish a fund that supports matching grants, whereby recipients of supply-side support programs cover some share of the total cost of the support. In some countries, matching grant or "catalytic funds" are also used to promote collaboration between foreign investors and groups of local firms (or farms), or between foreign investors and domestic institutions (for example, universities, research institutes, and so forth).

Establish sector forums for communication and coordination between government and the private sector around linkages and spillovers. One of the weaknesses of efforts to promote spillovers from FDI, particularly where a country has a limited number of large-scale investors, is that dialogue between government and investors tends to take place only on a bilateral basis. Making sustainable progress on spillovers, however, requires much more strategic and coordinated dialogue among government, investors (collectively, at least at the industry level), and other stakeholders. Indeed, getting investors (and the private sector more widely) to speak with a single voice is often one of the most valuable outcomes of such public-private dialogue. In the Ghanaian mining sector, for example, the Chamber of Mines emerged as a powerful body that has taken a lead role in promoting local economy linkages. And in the Sri Lankan garment sector, the Joint Apparel Association Forum has played a similarly instrumental role.

Multistakeholder partnerships can be effective in the design and delivery of linkage and spillover programs. Multistakeholder partnerships—often involving a combination of government, investors, donors, and/or technical experts (often NGOs)—have proven effective in designing and delivering linkage programs. Typically a special purpose vehicle, under a public-private partnership model, is established to actually run the programs. One example is the Source Trust (Armajaro), a nonprofit that brings together MNC chocolate manufacturers along with donors to improve traceability standards and deliver technical assistance to farmers. Another example is AgDevCo (a donor-funded private equity firm focusing on sustainable agriculture in developing countries), which teams up with Rio Tinto and the governments of Great Britain and Mozambique to promote supplier development and linkages in Mozambique.

Multistakeholder dialogues can be effective in the management of expectations. Multistakeholder dialogues can help manage expectations on the distribution of rents, especially in extractive industries with a limited time horizon. The establishment and development of private-public dialogue (PPD) mechanisms can ensure that stakeholders feel adequately consulted and informed, not only about the policy formulation process, but also about the execution of the economic activities derived from FDI in extractive activities. PPDs provide a platform for communities to express their concerns directly to policy makers and to manage potential conflicts with foreign investors that can have the potential to undermine the benefits of the investment.

Monitoring is critical both to ensure more effective policy and to encourage transparency and facilitate communications. Effective monitoring and evaluation of linkage programs has been lacking in most developing countries. Rigorous monitoring is important for a number of reasons. First, it will ensure more effective development, targeting, and delivery of interventions to promote spillovers. Second, it offers an opportunity to strengthen local civil society and NGOs in monitoring things like health, safety, and environmental compliance. Third, the transparency it brings with it will facilitate improved stakeholder relations. And finally, and perhaps most importantly, it can be used objectively to ensure that investors are delivering on their commitments (for example, through publishing local spending and other measures of local spillovers).

Notes

1. In Australia, a country with a successful linkage/spillover policy and impact in place, the development of a company's local procurement plan is a condition to receiving certain incentives. Norway developed a strong local service and construction sector related to oil exploration and development using a combination of local content policies and other support of the local supply industry, which is monitored by the Goods and Services Office established in 1972 (Tordo et al. 2013).
2. Tordo et al. (2013), for example, describe different local content strategies and objectives, performance indicators, and tools that encourage the development of linkages in the oil and gas sector.

References

Tordo, S., M. Warner, O. Manzano, and Y. Anouti. 2013. *Local Content Policies in the Oil and Gas Sector*. World Bank Study 78994, Washington, DC: World Bank. Available at: http://www-wds.worldbank.org/external/default/WDSContentServer/WDSP/IB/2013/07/12/000445729_20130712141852/Rendered/PDF/789940REVISED000Box377371B00PUBLIC0.pdf.

World Bank. 2012. *Increasing Local Procurement by the Mining Industry in West Africa, Road Test Version*. Washington, DC: World Bank.

Environmental Benefits Statement

The World Bank Group is committed to reducing its environmental footprint. In support of this commitment, the Publishing and Knowledge Division leverages electronic publishing options and print-on-demand technology, which is located in regional hubs worldwide. Together, these initiatives enable print runs to be lowered and shipping distances decreased, resulting in reduced paper consumption, chemical use, greenhouse gas emissions, and waste.

The Publishing and Knowledge Division follows the recommended standards for paper use set by the Green Press Initiative. Whenever possible, books are printed on 50 percent to 100 percent postconsumer recycled paper, and at least 50 percent of the fiber in our book paper is either unbleached or bleached using Totally Chlorine Free (TCF), Processed Chlorine Free (PCF), or Enhanced Elemental Chlorine Free (EECF) processes.

More information about the Bank's environmental philosophy can be found at http://crinfo.worldbank.org/wbcrinfo/node/4.